Earthquake Engineering:
Advanced Concepts and Mechanisms

Earthquake Engineering: Advanced Concepts and Mechanisms

Editor: Courtney Silver

R CALLISTO REFERENCE

www.callistoreference.com

Callisto Reference,
118-35 Queens Blvd., Suite 400,
Forest Hills, NY 11375, USA

Visit us on the World Wide Web at:
www.callistoreference.com

ISBN: 978-1-64116-183-1 (Hardback)

Cataloging-in-Publication Data

Earthquake engineering : advanced concepts and mechanisms / edited by Courtney Silver.
    p. cm.
Includes bibliographical references and index.
ISBN 978-1-64116-183-1
1. Earthquake engineering. 2. Earthquake resistant design.
3. Structural dynamics. I. Silver, Courtney.
TA654.6 .E27 2019
624.176 2--dc23

Table of Contents

Preface

The purpose of the book is to provide a glimpse into the dynamics and to present opinions and studies of some of the scientists engaged in the development of new ideas in the field from very different standpoints. This book will prove useful to students and researchers owing to its high content quality.

Earthquake engineering is a field of engineering, which includes designing, and analyzing structures with respect to seismic loading. The main goal of earthquake engineering is to make structures, which are more earthquake resistant and resilient. The field is rapidly evolving with a major focus on protecting the society, and the environment by minimizing the seismic risk to socially and economically acceptable levels. The field includes concepts like seismic design, failure mode, earthquake-resistant construction, etc. Seismic vibration control devices are used in building structures to reduce the potential impact of earthquakes. This book includes contributions of experts and scientists, which will provide innovative insights into this field. It studies, analyzes and upholds the pillars of earthquake engineering and its utmost significance in modern times. Students, researchers, experts and all associated with the study of earthquake engineering will benefit alike from this text.

At the end, I would like to appreciate all the efforts made by the authors in completing their chapters professionally. I express my deepest gratitude to all of them for contributing to this book by sharing their valuable works. A special thanks to my family and friends for their constant support in this journey.

Editor

Intelligent Seismic-Acoustic System for Identifying the Area of the Focus of an Expected Earthquake

Telman Aliev

Additional information is available at the end of the chapter

Abstract

Over the last few years, the following theoretical and practical research, technologies and algorithms have been developed allowing one to determine characteristics of noise contained in noisy seismic-acoustic signals. These characteristics (noise variance, cross-correlation function between the useful signal and the noise, relay estimations, etc.) are used to indicate the start of anomalous seismic processes (ASPs) as the earthquake preparation process. Using these characteristics, technologies for determining informative attributes of identification of the latent period of origin of ASPs have been developed. Based on those technologies, stations for robust noise monitoring of ASPs have been created and are currently functioning in Azerbaijani. Noise monitoring of ASPs was conducted from 2010.07.01 to 2014.06.01 on nine such stations built at wells of varying depth. Based on the results of obtained experimental data, an intelligent system has been built. It allows identifying the location of the area of an earthquake 10–20 h in advance, using combinations of time of change in the estimate of correlation function $R_{X\varepsilon}(\mu)$ between useful signal $X(i\Delta t)$ and noise $\varepsilon(i\Delta t)$ of seismic-acoustic information received from different stations. In the long term, the system can be used by seismologists as a tool for determining the location of the area of an expected earthquake.

Keywords: seismic-acoustic signal, anomalous seismic process, informative attributes, earthquake focus, robust noise monitoring, intelligent system, correlation function

1. Introduction

Nature and origin of earthquakes are a subject of many research papers [1]. There is also a lot of material devoted to obtaining seismic information from the earth's deep layers [2, 3]. Seismic signals received during earthquakes are analysed by means of noise analysis [4–8], designing continuous monitoring system [9], earthquake damage assessment and damage minimization [10–12], wavelet transform and finite elements [13–15]. Earthquake prediction-related problems are treated as a primary trend of research [16–19]. Various means and tools have been and are being developed and commissioned [20–22], including earthquake early warning systems for general population, models and technologies for prompt response of rescue groups [23–27]. Despite all the efforts and achievements, we still fail to predict earthquakes early enough, the results of which can be catastrophic indeed.

The authors of references [9, 28, 29] suggest employing a seismic acoustic system to monitor earthquake origin. Such a system comprises nine stations performing robust noise monitoring of anomalous seismic processes (RNM ASP) as a single network. Experimenting on those stations carried out from July 2010 demonstrated that incipient ASPs lead to the appearance of a cross-correlation between noise and useful signal of seismic acoustic data.

Using the varying estimate of the cross-correlation function between useful signal and noise, each of the stations in the network separately is reliable enough to indicate the incipient ASP processes preceding earthquakes. Nevertheless, the accuracy of coordinates of a coming earthquake determined by means of those stations proves to be insufficient. However, it has been established by way of experiments that we can design an intelligent neural network system that would use these stations for locating the ASP area. The following is our thoughts on how to build the said system.

2. Problem statement

In the regions of high seismic activity, incipient ASPs usually cause an earthquake to occur after the normal seismic state period T_0, as period T_1 ends.

T_0 and T_1 have different duration, but to monitor the start of the ASP origin, we basically need a reliable indicator of the start of T_1, which is discussed in references [4, 9, 28].

In reference [9], the authors propose creating a technology and a system for registering the starting point of T_1. According to the experiments laid out in references [4, 9, 28], however, T_1 does not start only during ASP origin. Therefore, apart from registering the start of T_1, to monitor the start of incipient ASPs, we also need the changes in the estimate of the cross-correlation function $R_{X\varepsilon}(\mu)$ between the useful signal $X(i\Delta t)$ and the noise $\varepsilon(i\Delta t)$ to be indicated.

In the following paragraphs, we will use the estimate $R_{X\varepsilon}(\mu)$ of the seismic-acoustic signal $g(i\Delta t)$ as an informative attribute to indicate the beginning of the ASP origin. To do that, we must calculate $R_{X\varepsilon}(\mu)$ while monitoring.

Further on, practical use of the RNM ASP network also requires a technology for finding the location of the earthquake zone. To this end, we must first review the known methods designed to calculate the focus of an earthquake [30, 31] based on seismic data acquired by means of regular above-ground stations.

In that case, we find the focus of an earthquake from the difference between the amounts of time it takes P and S waves to reach each of the above-ground stations. The velocity of P wave propagation is higher than that of S wave propagation. The velocity of P wave in a homogeneous isotropic medium is

$$v_P = \sqrt{\frac{k + \frac{4}{3}\mu}{\rho}},$$ (1)

where k is the volume factor, μ is the shear modulus and ρ is the density of the medium that waves penetrate.

The velocity of S wave propagation is

$$v_S = \sqrt{\frac{\mu}{\rho}},$$ (2)

where μ is the shear modulus and ρ is the density of the material penetrated by the waves.

The distance from a regular above-ground seismic station to the focus is determined by multiplying the time difference by the difference in velocity:

$$S = \Delta T (v_p - v_s).$$ (3)

After the distance between the epicentre and the different seismic stations has been determined, the coordinates of the focus are found geometrically. Unfortunately, in all known cases, the coordinates of epicentres and hypocentres in seismic monitoring systems are determined after actual earthquakes.

Our experimental research showed that, for many reasons, it is practically impossible to use the obtained results to calculate the coordinates of the ASP areas on RNM ASP stations by means of the said technology.

Therefore, the present paper poses the problem of developing an intelligent neural network system for monitoring the ASP origin, identifying the location of the area and determining the approximate magnitude of an anticipated earthquake.

3. Determining the informative attributes of the hidden period of ASP origin

As ASP emerges at the start of T_1, the first estimates to change are those of cross-correlation function $R_{X\varepsilon}(\mu = 0)$ between useful signal $X(i\Delta t)$ and noise $\varepsilon(i\Delta t)$, variance D_ε of the noise and noise correlation $R_{X\varepsilon\varepsilon}(\mu = 0)$ [4, 9, 28]. This happens because noise$\varepsilon(i\Delta t)$ is formed due to the effects of the incipient ASP as period T_0 begins. Consequently, in T_1, a correlation emerges between useful signal $X(i\Delta t)$ and noise $\varepsilon(i\Delta t)$, causing an abrupt increase in the estimate $R_{X\varepsilon}(\mu)$. Therefore, we can consider $R_{X\varepsilon}(\mu)$ the main informative attribute and use it while monitoring the hidden period of ASP origin.

Starting from July 2010, we used traditional technologies as well as robust noise technologies on RNM ASP stations to detect the start of the hidden period of ASP origin. A sufficiently reliable registration of period T_1 by means of estimates obtained through traditional spectral and correlation technologies proved to be unattainable. The use of robust noise technology, however, caused an abrupt change in the estimate of the cross-correlation function $R_{X\varepsilon}(\mu)$ at the start of T_1. That was a crucial factor, adding to the validity of the monitoring. Considering that, we used $R_{X\varepsilon}(\mu)$ as an informative attribute in the monitoring of ASP origin, while creating the RNM ASP network.

The relay correlation function $R^*_{gg}(\mu = 0)$ between useful signal $X(i\Delta t)$ and noise $\varepsilon(i\Delta t)$ is

$$R^*_{X\varepsilon}(\mu = 0) \approx \frac{1}{N} \sum_{i=1}^{N} \left[\text{sgn } g(i\Delta t)g(i\Delta t) - 2\,\text{sgn } g(i\Delta t)g((i+1)\Delta t) + \text{sgn } g(i\Delta t)g((i+2)\Delta t) \right]. \quad (4)$$

Knowing the estimates $R^*_{X\varepsilon}(\mu = 0), R^*_{gg}(\mu = 1), R_{gg}(\mu = 1)$ and considering the equality relationship between $R^*_{gg}(\mu = 1)$ and $R_{gg}(\mu = 1)$ and $R^*_{X\varepsilon}(\mu = 0)$ and $R_{X\varepsilon}(\mu = 0)$

$$\frac{R^*_{gg}(\mu = 1)}{R_{gg}(\mu = 1)} = \frac{R^*_{X\varepsilon}(\mu = 0)}{R_{X\varepsilon}(\mu = 0)}, \quad (5)$$

we can calculate $R_{X\varepsilon}(\mu = 0)$ from this formula:

$$R_{X\varepsilon}(\mu = 0) = \frac{R_{gg}(\mu = 1) R^*_{X\varepsilon}(\mu = 0)}{R^*_{gg}(\mu = 1)}. \quad (6)$$

We were able to conclude from our experiments that to achieve more trustworthy monitoring results, we should also use the estimates of noise correlation $R_{X\varepsilon\varepsilon}(\mu = 0)$ and noise variance D_ε as extra informative attributes. Those estimates are calculated from the following expressions [4, 9, 28]:

$$R_{x\varepsilon\varepsilon}(\mu) = R_{x\varepsilon}(\mu) + D_\varepsilon = \frac{1}{N}\sum_{i=1}^{N}\left[g^2(i\Delta t) + g(i\Delta t)g((i+2)\Delta t) - 2g(i\Delta t)g((i+1)\Delta t)\right] \qquad (7)$$

$$D_\varepsilon = R_{X\varepsilon\varepsilon}(\mu = 0) - R_{X\varepsilon}(\mu = 0). \qquad (8)$$

As we can see, $R^*_{X\varepsilon}(\mu = 0)$, $R_{X\varepsilon}(\mu = 0)$, $R_{X\varepsilon\varepsilon}(\mu = 0)$ and D_ε are determinable from Eqs. (4), (6), (7) and (8). These estimates raise the validity of ASP monitoring to an adequate level.

4. Technology and systems for locating the ASP origin area

An earthquake takes place as soon as an ASP hits a critical point of development. Earthquake magnitude and the radius of its focus are contingent on the structure and nature of the strain-stress distribution in the rocks in a particular location. A jump-like rock deformation emits elastic waves. The amount of the deformed mass is a significant aspect that determines the intensity of the shock and the formation of noise $g(i\Delta t)$. Core bursts follow periods T_1 of earthquake preparation that can be as long as dozens of hours.

Analysing the seismic data from the acoustic sensors at suspended oil wells, we find that as ASPs start, seismic-acoustic noise travelling in the earth's deep layers anticipates the earthquake by dozens of hours [4, 9, 28]. Experiments show that RNM ASP stations can adequately monitor the beginning of T_1 by the above-described technology (**Figure 1**). Further on, we will consider working out an intelligent technology for locating the ASP area, using the data from the stations installed in nine seismically active regions of the Caspian Sea (**Figure 1**). The geographical coordinates and well depths of the stations are given in **Table 1**.

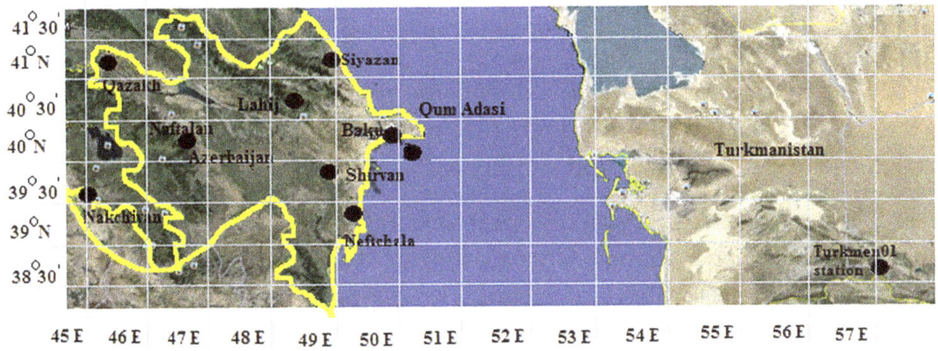

Figure 1. Map of the locations of RNM ASP stations in the seismically active region of the Caspian Sea.

No.	Station	Latitude	Longitude	Well depth	Start of operation
1	Qum Island	40.310425°	50.008392°	3500 m	July 2010
2	Siazan	41.046217°	49.172058°	3145 m	November 2011
3	Naftalan	40.609521°	46.791458°	4000 m	May 2012
4	Shirvan	39.933170°	48.920745°	4900 m	November 2011
5	Neftchala	39.358333°	49.246667°	1430 m	June 2012
6	Nakhchivan	39.718000°	44.876000°	1800 m	March 2013
7	Qazakh	41.311889°	45.108611°	200 m	August 2013
8	Turkmenistan	38.530089°	56.654472°	300 m	August 2013
9	Cybernetic	40.223252°	49.800833°	10 m	February 2014

Table 1. Geographical coordinates and well depths of RNM ASP stations.

According to the results of the experiments on the RNM ASP stations (**Figure 2**), the seismic noises caught by hydrophones from the earth's deep layers are immediate precursors of earthquakes.

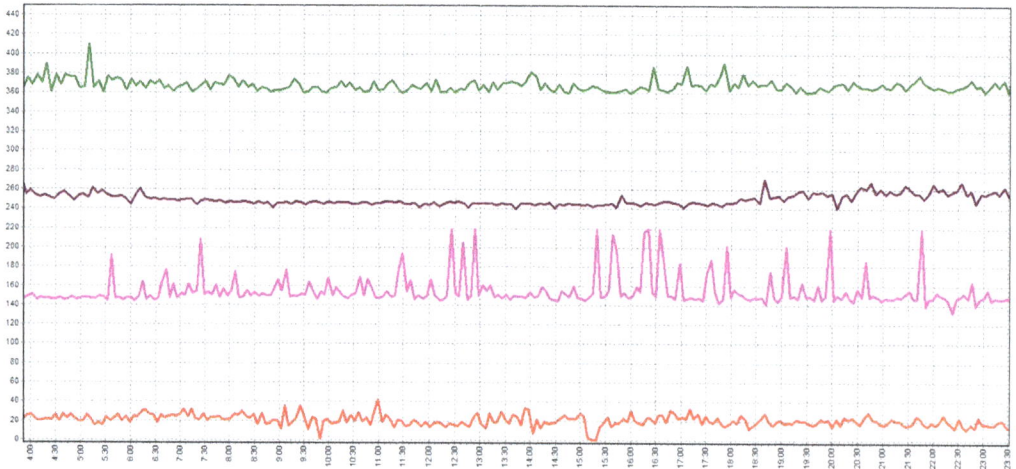

Figure 2. Siazan, Qum Island, Shirvan, Neftchala: 2013-03-26 Georgia-Russia.

Those noises were measured and analysed, and the relevant data was forwarded from the stations to the server of the monitoring centre (MC) on a high-speed radio channel via satellite. The received data can also be forwarded to other MCs in other countries of a particular region.

As seen from **Figure 1**, RNM ASP stations were put into operation one by one starting from July 2010: first at Qum Island, then in Shirvan, Siazan, Naftalan, Neftchala, Nakhchivan (on the borders with Turkey and Iran), Turkmen01 (in Turkmenistan), Qazakh (on the Georgian border) and Cybernetic (in Baku). The last three stand on 300-, 200- and 10-m deep water wells, respectively. Pipes in the wells naturally fill with water. Hydrophones were installed inside the pipes at 10–20 m from the water level. The built network of stations allowed us to conduct large-scale experiments that have demonstrated that the seismic-acoustic noises emerging during ASP origin spread within a 300–500 km radius many hours before the seismic waves can be detected by above-ground stations.

The operation of the network involves synchronous robust analysis of seismic-acoustic signals received from all stations. Values of noise parameters $R_{x\varepsilon}(\mu), R_{x\varepsilon\varepsilon}(\mu), D_{\varepsilon}$ are sent to the MC from the stations (**Figure 2**). By the changes in those estimates, the starting points T_{1i} and T_{1j} of ASP origin are indicated for the ith and jth stations, respectively.

We have established that each RNM ASP station separately can adequately indicate an incipient ASP, as well as that we can use the results obtained during the operation of the network as a basis to create an intelligent technology for locating the zone of an anticipated earthquake. For this purpose, the network first determines the combinations of indication moments T_{1i} and T_{1j}. Those combinations, together with the geographical coordinates of the stations are the source data used to locate the ASP origin area. For the results to be adequate and trustworthy, it is appropriate that, in addition to the combinations of indication moments, time differences $T_{1i} - T_{1j}$ should also be used for each chosen pair of stations. That is, the combination T_{1i}, T_{1j} alone is insufficient as source data; we also need to determine the difference in time of ASP indication between the stations $\Delta\tau_{ij} = \left(T_{1i} - T_{1j} \right)$.

It is not easy to accurately identify the start of the time of indication T_{1i} by means of the estimates of noise parameters. For this reason, our system duplicates the process of determining $\Delta\tau_{ij}$, when the time difference $\Delta\tau ij = (T_{1i} - T_{1j})$ is also determined, using the extreme value of cross-correlation function $R_{g_ig_j}(\mu_{max})$ between the signals $g_i(i\Delta t)$ and $g_j(i\Delta t)$ obtained from different combinations of stations. The following expressions are used for this purpose:

$$R_{g_ig_j}(\mu_{max}) = \frac{1}{N}\sum_{i=1}^{N} g_i(i\Delta t)g_j(i+\mu)\Delta t \tag{9}$$

$$R_{g_ig_j}^*(\mu_{max}) = \frac{1}{N}\sum_{i=1}^{N} g_i^2(i\Delta t)g_j^2(i+\mu)\Delta t \tag{10}$$

$$R_{g_ig_j}^*(\mu_{max}) = \frac{1}{N}\sum_{i=1}^{N} g_i(i\Delta t)g_j^2(i\Delta t) \tag{11}$$

Then the difference in the time of indication between different stations on the server of the monitoring centre is determined in the following order:

1.   Finding the time of registration of the start of period T_{1i} of ASP origin by the first station Qum Island.

2. Finding the time of registration for each subsequent station (Shirvan, Siazan, Naftalan, etc.).

3. Finding the sets of estimates of cross-correlation functions $R_{g_v g_j}(i\Delta t)$, $R_{g_v g_j}(i\Delta t)$ by Eqs (9)–(11) and, choosing from the results the time shifts $\mu \cdot \Delta t$, at which the curve of the cross-correlation function has the peak value, i.e. the extreme value; using those time shifts to determine $\Delta \tau_{ij} = \left(T_{1i} - T_{1j} \right)$.

4. The found time differences $\Delta \tau_{1i} = \left(T_{1i} - T_{1j} \right)$ are used as source data to locate the ASP area.

We see that in our system (**Figure 2**), the values of the noise parameters $R_{X\varepsilon}(\mu)$, $R_{X\varepsilon\varepsilon}(\mu)$ and D_ε obtained by the RNM ASP stations are synchronously sent via satellite communication to the MC server. On the basis of the results, combinations of sequences of indication times T_{1i} T_{1j} and combinations of time differences $\Delta \tau_{ij}$ are formed and then used as source data in locating of the earthquake area.

Our long-term experiments on the stations were conducted from July 2010 to June 2014. They identified the following 13 seismically active zones in Azerbaijan and nearby regions within a 500–600 km radius around the network of the RNM ASP stations.

I. Turkmen coast of the Caspian Sea;

II. South of the Absheron peninsula (in the Caspian Sea);

III. North of the Absheron peninsula (in the Caspian Sea);

IV. Shirvan (region of Azerbaijan);

V. North-western regions of Azerbaijan;

VI. Southern regions of Azerbaijan;

VII. South of the Caucasus region of the Russian Federation;

VIII. North-eastern regions of Iran;

IX. North-western regions of Iran (near Tabriz);

X. Iranian-Iraqi-Turkish border;

XI. Northern regions of Iran;

XII. Eastern regions of Turkey;

XIII. Western regions of Georgia (Black Sea).

We have previously given some of the results obtained by means of the RNM ASP stations in those zones in [9].

Those 13 zones have experienced many earthquakes with magnitude 3–4 in the last 1.5–2 years. Combinations of the sequence of the times of the ASP indication by Qum Island, Shirvan, Siazan, Neftchala, Naftalan and Nakhchivan stations for each of them practically overlapped. Analysing the records, we have concluded that each combination of the time of corresponds to one specific zone. After 2 years of working at the interpretation of the results of our experiments, we were able to accurately locate the zone of an expected earthquake instinctively, using those combinations. Realising that earthquake areas could be located with the help of expert systems (ESs), we established that it was possible to design an ES for seismologists to use a network of the RNM ASP stations as a toolkit in locating the area of anticipated earthquakes.

The foundation of the proposed experimental version of such an ES for locating the ASP area (ESILA) is the knowledge base (KB) consisting of sets W_1, W_2, W_3, ..., W_{13} of the locations of the respective areas. Elements of each set are built from the data in the charts containing the parameters of all earthquakes registered by the stations in the mentioned 13 areas from July 2010 up to this day. Elements of the base comprise the combination of the sequence of times T_{1i}, T_{1j} when ASP was registered, the combination of the differences in times of the indication $\Delta\tau_{ij}$, and the combination of the estimates of the cross-correlation function $R_{X\varepsilon}(\mu = 0)$. They also contain the value of magnitude M_i found during respective earthquakes by above-ground seismic stations, as well as earthquake date. If only one element is available, the KB has the following form:

$$
\begin{aligned}
W_1 &\left\{ \begin{bmatrix} T_{11}^{1(1)} & T_{11}^{2(1)} & \cdots & T_{11}^{6(1)} & M_1 \\ \Delta\tau_{11}^{1} & \Delta\tau_{21}^{1} & \cdots & \Delta\tau_{61}^{1} & M_1 \\ R_{X\varepsilon}^{1(1)}(\mu=0) & R_{X\varepsilon}^{2(1)}(\mu=0) & \cdots & R_{X\varepsilon}^{6(1)}(\mu=0) & M_1 \end{bmatrix} \right. \\[2mm]
W_2 &\left\{ \begin{bmatrix} T_{11}^{1(2)} & T_{11}^{2(2)} & \cdots & T_{11}^{6(2)} & M_2 \\ \Delta\tau_{11}^{2} & \Delta\tau_{21}^{2} & \cdots & \Delta\tau_{61}^{2} & M_2 \\ R_{X\varepsilon}^{1(2)}(\mu=0) & R_{X\varepsilon}^{2(2)}(\mu=0) & \cdots & R_{X\varepsilon}^{6(2)}(\mu=0) & M_2 \end{bmatrix} \right. \\[2mm]
&\qquad\qquad\qquad\qquad \vdots \\[2mm]
W_{13} &\left\{ \begin{bmatrix} T_{11}^{1(13)} & T_{11}^{2(13)} & \cdots & T_{11}^{6(13)} & M_{13} \\ \Delta\tau_{11}^{13} & \Delta\tau_{21}^{13} & \cdots & \Delta\tau_{61}^{13} & M_{13} \\ R_{X\varepsilon}^{1(13)}(\mu=0) & R_{X\varepsilon}^{2(13)}(\mu=0) & \cdots & R_{X\varepsilon}^{6(13)}(\mu=0) & M_{13} \end{bmatrix} \right.
\end{aligned}
\tag{12}
$$

Sets $W_1 - W_{13}$ of the experimental KB consist of dozens of elements and are updated with new ones during new earthquakes. After the monitoring and registration of the time of the start of a current ASP, the stations build current combinations of the sequence of the indication times

T_{1i}, T_{1j}, the combination of differences in the indication times $\Delta\tau_{ij}$, and the combination of estimates $R_{X\varepsilon}(\mu)$.

In January 2014, the experiments on locating the earthquake areas by means of ESILA started. The procedure is as follows. The monitoring results obtained by the RNM ASP network are used to form a current element, which is compared with those in the sets W_1, W_2, W_3, ..., W_{13} within the identification unit of the ES (IUES). Should there be any match, the zone of an earthquake is located based on the order number of the current element. The number of the zone is memorised in the ES decision-making unit (DMU) and the current element is saved in the set in the KB. In this manner, more and more elements are continuously saved into the KB while ESILA is functioning. The RNM ASP network and ESILA work as a unified system.

ESILA was tested during all subsequent earthquakes to confirm the adequacy and validity of the results it produces. It became obvious that it is really possible to practically use this experimental version to locate ASP zones. Therefore, the system can be useful for determining the areas of anticipated earthquake. With this in mind, the features of the DMU of ESILA were updated with the feature of compiling the following types of information and presenting it to seismologists:

1. Date of current ASP, the number of area of anticipated earthquake;

2. Current monitoring results from RNM ASP stations;

3. Assessed lead time at the start of ASP monitoring compared with the time of indication by above-ground stations;

4. Elements registered in the relevant set during the previous ASP in the assumed earthquake area (including dates);

5. The number of elements identical to the current ones;

6. Magnitudes of previous earthquakes;

7. Minimum magnitude of anticipated earthquake; and

8. If KB contains no elements identical at least to some elements in the sets W_1–W_{13}, DMU gives out the information that locating the earthquake area is impossible.

5. Technology for determining the approximate value of magnitude of an expected earthquake using a neural network

The analysis of the results of the experimental identification of the location of the ASP area has demonstrated that, with the current estimates $R_{X\varepsilon}(\mu)$, $R_{X\varepsilon\varepsilon}(\mu)$, D_ε and knowing the distance from the area to the RNM ASP stations, it is possible to determine the approximate value of the minimum magnitude of an expected earthquake using a neural network. Research shows that neural networks can be used for this purpose [32, 33]. It was found appropriate to use the information contained in the sets W_1–W_{13} to train neural networks. The block diagram of the neural network ($N3 = 1$) is functioning in the following way. The content of the corresponding elements of the sets W_1, W_2, W_3, ..., W_{13} is transmitted to the outputs X_1, X_2, ..., $X_{N\,1}$ of the neuron, i.e. the combinations of times of the ASP indication T_{1ij}, differences of indication time $\Delta\tau_{ij}$ and the estimate $R_{X\varepsilon}(\mu = 0)$ are received at the inputs of the neuron one by one; the magnitude M_i of the earthquake registered by ground stations is established at the output of the neuron. The training is carried out successively from earthquake area I to earthquake area XIII. For instance, during the training of the neuron on area III, i.e. during the earthquake with the area in the Caspian Sea, the monitoring results obtained at the stations in Siazan, Qum Island, Neftchala and Turkmen01 (Turkmenistan) are successively transmitted from the KB to the inputs of the neuron. The value of the magnitude $M3$ is given to the output. During the training of the neuron to determine the magnitude in area XII, i.e. in East Turkey, then the monitoring data of Qazakh, Naftalan, Shirvan and Nakhchivan are sent to the input of the neuron and the magnitude $M12$ goes to the output. Thus, the parameters of the ASP previously registered by the RNM ASP stations are used for the neural network training. At the same time, the coordinates of the location of the earthquake areas are used in the DMU to determine the approximate distance S_I between the stations and the areas, which are also transmitted to the inputs of the neural network. Based on the source data written in the elements of the sets W_1–W_{13} and the distances S_1–S_9 from the ASP area to each station, the neural network learns to determine the approximate magnitude of an expected earthquake. Owing to this, after the training stage and in the process of the current monitoring of the ASP, when the current combinations of corresponding estimates are transmitted to the neuron outputs, the code of the corresponding magnitude M of the expected earthquake forms on the output y_3 [1]. The result is sent to the input of the DMU of ESILA.

During the operation of the neural network and the ES, every time the coordinates and approximate magnitude of every expected earthquake have been identified, the obtained results are compared with the coordinates and magnitude of actual earthquakes registered by ground seismic stations. The obtained difference is further used to correct the KB and in the training of the neural network. Therefore, the KB is improved in the course of time, with the training level of the neural network constantly improving. This results in increased reliability, authenticity and adequacy of identification of the location and magnitude of expected earthquakes.

Analysing the experience of the use of the ES in identifying the location of the area of the expected earthquake and of the neural network in determining its magnitude, we have established that to enhance the trustworthiness of the results we must increase the number of RNM ASP stations in the network. To this end, the Nakhchivan station near the border with Turkey and Iran and Turkmen01 station in Turkmenistan (**Figure 1**) were commissioned. In July 2013, the Qazakh station and Cybernetic station were built on the Georgian border (**Figure 1**) and at the Institute of Control Systems (Baku), respectively.

Experimental monitoring performed by the Cybernetic station installed in the basement of the Institute of Control Systems at a 10-m deep well gave the results that matched the readings of the Qum Island station standing at a 3500-m deep well.

6. Results of experiments on locating earthquake areas (January 2013–July 2014)

After the test operation of the system started in January 2013, certain identification errors were registered during weaker earthquakes. Errors in identification were also registered on two or three stations simultaneously due to a malfunction of the power lines, communication and hydrophone, controller and other units. No errors were detected for earthquakes with strength over five points, when the stations were operating normally.

Due to the length of the list of all identified locations of anticipated earthquakes for 2013–2014, we included in **Table 2** below only 10 locations for earthquakes with magnitude over five from the period from January 2013 to July 2014. **Figures 2–11** show the charts of ASPs that preceded those earthquakes. The first column of the table references data taken from the Euro-Mediterranean Seismological Centre (EMSC) website (http://www.emsc-csem.org/#2).

The time of the earthquakes in the table is in UTC, while the time in the charts is local (Baku time, UTC + 4).

Identified locations of anticipated earthquakes can be found in column 22. For authenticity, each row of the table is supported by a respective chart (**Figures 2–11**) drawn by the RNM ASP stations in the process of origin of a respective ASP.

Sign '*' implies that the response of a station to the incipient ASP of an anticipated earthquake was weak, sign '−' that the registered estimate of $R_{\chi\varepsilon}(\mu)$ is beneath the threshold value.

In Row 1 of **Table 2**, we see the results of the identification of location for the Georgian earthquake of 26 March 2013. **Figure 2**, in turn, shows that the earthquake start was registered at 04:15 by the Siazan station, at 04:30 by the Qum Island station, at 06:50 by the Shirvan station and at 08:30 by the Neftchala station. Naftalan station was not functioning at the time of that earthquake. Nevertheless, the system concluded that such manner of ASP registration corresponded to area VII. The ASP was registered 8–10 h before the earthquake occurred.

Row 2 contains the results of the identification of location for the Georgian earthquake of 28 May 2013. From the chart we can see that Siazan, Naftalan, Shirvan and Qum Island stations indicate the ASP origin over 20 h before the earthquake (even without the data from the malfunctioning Naftalan station). The northern (Siazan) and north-western (Qum Island) stations were the first to detect an anomaly. That being said, the RNM ASP stations indicated the start of ASPs in the following order: 07:30 — Naftalan; 09:10 — Siazan; 09:45 — Shirvan; 11:40 — Qum Island (**Figure 3**). The system located the zone of the earthquake by 18:00 (Baku time), i.e. 10–11 h before the above-ground stations.

1	2	4	5	6	7	8	9	10	11	12
No.	Date, time, coordinates, magnitudes and depth of earthquake epicentre	$\Delta\tau_{12}$	$\Delta\tau_{13}$	$\Delta\tau_{14}$	$\Delta\tau_{15}$	$\Delta\tau_{16}$	$\Delta\tau_{17}$	$\Delta\tau_{18}$	$\Delta\tau_{19}$	$R_{X\varepsilon}$
1	2013-03-26, 23:35:25.0 UTC, 43.19 N; 41.67 E, mag 4.8, 10 km	35	−120	−	135	*	−	*	−	300
2	2013-05-28, 00:09:52.0 UTC, 43.22 N; 41.58 E, mag 5.2, 2 km	−115	−150	−250	−	*	−	*	−	150
3	2013-09-17, 04:09:13.0 UTC, 42.13 N; 45.80 E, mag 5.1, 2 km	−	−150	−	390	*	120	*	−	100
4	2013-11-24, 18:05:41.0 UTC, 34.06 N; 45.52 E, mag 5.6, 10 km	−	*	−	−10	−60	*	*	−	160
5	2014-01-10, 00:45:31.0 UTC, 41.86 N; 49.41 E, mag 4.8 80 km	−	20	−	110	*	−	−10	−	110
6	2014-01-14, 13:55:02.0 UTC, 40.33 N; 52.95 E, mag 5.2, 48 km	−	−45	*	−120	*	*	−	−	160
7	2014-01-28, 23:47:35.0 UTC, 32.45 N; 50.02 E, mag 4.9, 10 km	−135	−	−	100	−300	−	−	−	120
8	2014-02-10, 12:06:48.0 UTC, 40.23 N; 48.63 E, mag 5.4, 55 km	−300	−	−	−	45	75	−	−	75
9	2014-06-07, 06:05:32.4 UTC, 40.32 N; 51.58 E, mag 5.4, 44 km	145	20	−	−70	−	−	−	120	80
10	2014-06-29, 17:26:10.4 UTC, 41.62 N; 46.68 E, mag 5.1, 20 km	305	−85	−	−	−	315	*	−	100

13	14	15	16	17	18	19	20	21	22
No.	$R_{2X\varepsilon}$	$R_{3X\varepsilon}$	$R_{4X\varepsilon}$	$R_{5X\varepsilon}$	$R_{6X\varepsilon}$	$R_{7X\varepsilon}$	$R_{8X\varepsilon}$	$R_{9X\varepsilon}$	Number and location of the area of expected earthquake
1	50	100	–	140	–	–	–	–	Georgia (Sak'art'velo)
2	150	160	250	–	–	–	–	–	Georgia (Sak'art'velo)
3	–	40	–	80	–	80	–	–	Caucasus Region, Russia
4	–	–	–	150	250	–	–	–	Iran-Iraq Border
5	–	110	–	110	–	–	40	–	Caspian Sea, Offshore Azerbaijan
6	–	100	–	120	–	–	–	–	Turkmenistan
7	110	–	–	–	180	–	–	–	Western Iran
8	130	–	–	–	260	230	-	-	Azerbaijan
9	25	40	–	100	–	–	–	80	Offshore Turkmenistan
10	20	120	–	–	–	25	–	–	Azerbaijan

Table 2. Identified areas of expected earthquakes.

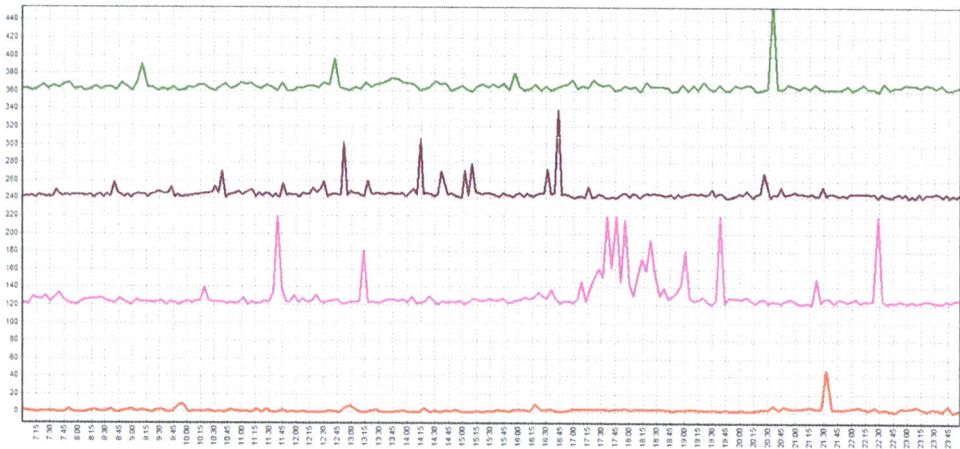

Figure 3. Siazan, Naftalan, Shirvan, Qum Island: 2013-05-27 Georgia (Sak'art'velo).

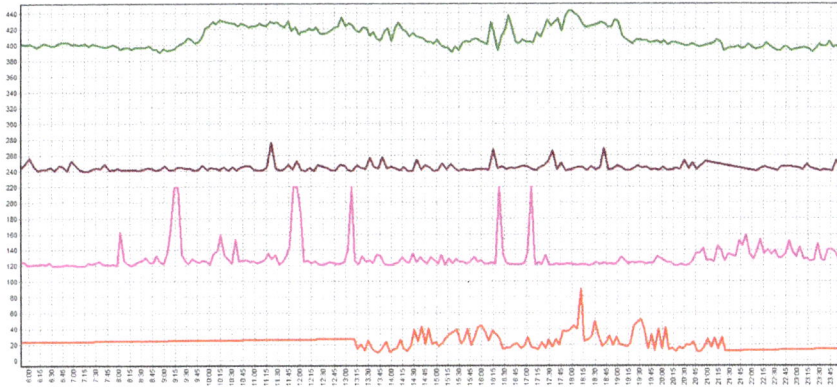

Figure 4. Qazakh, Siazan, Qum Island, Neftchala: 2013-09-16 Russia.

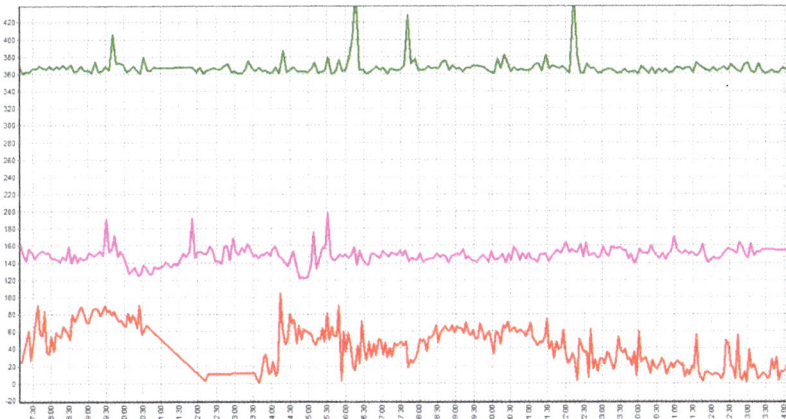

Figure 5. Qum Island, Neftchala, Nakhchivan: 2013-11-21 Iran-Iraq border.

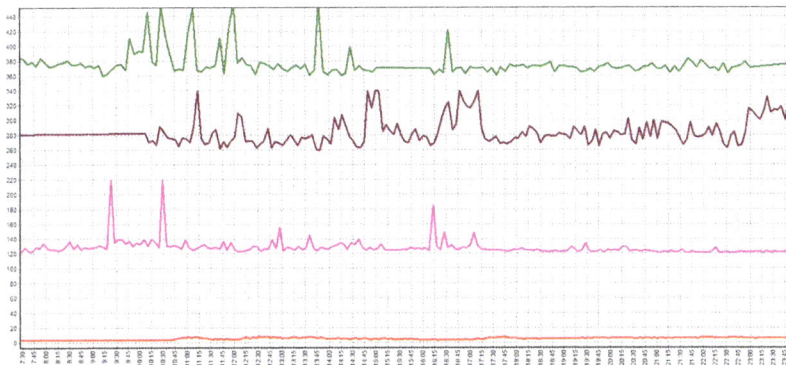

Figure 6. Siazan, Neftchala, Qum Island, Turkmen01: 2014-01-09 Caspian Sea, Offshore Azerbaijan.

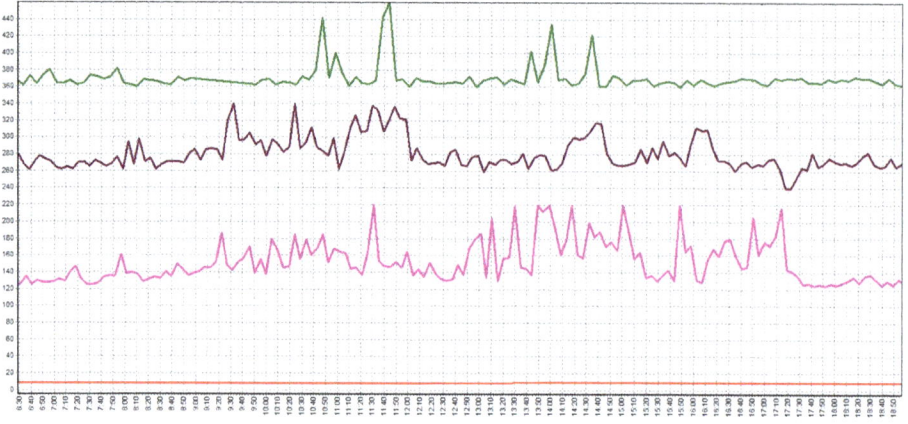

Figure 7. Siazan, Neftchala, Qum Island, Turkmen01: 2014-01-13 Turkmenistan.

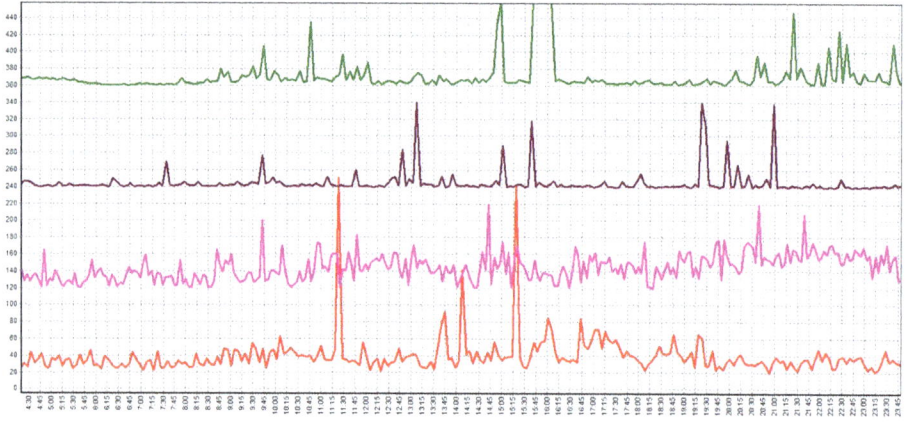

Figure 8. Qum Island, Shirvan, Nakhchivan, Neftchala: 2014-01-28 Western Iran.

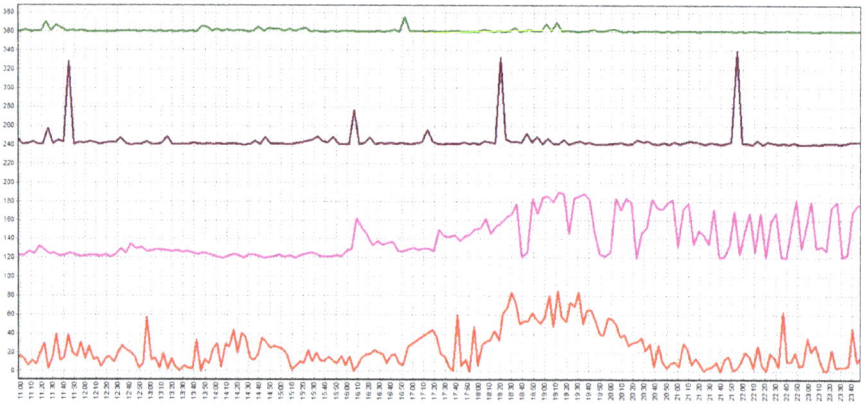

Figure 9. Qum Island, Shirvan, Qazakh, Nakhchivan: 2014-02-10 Azerbaijan.

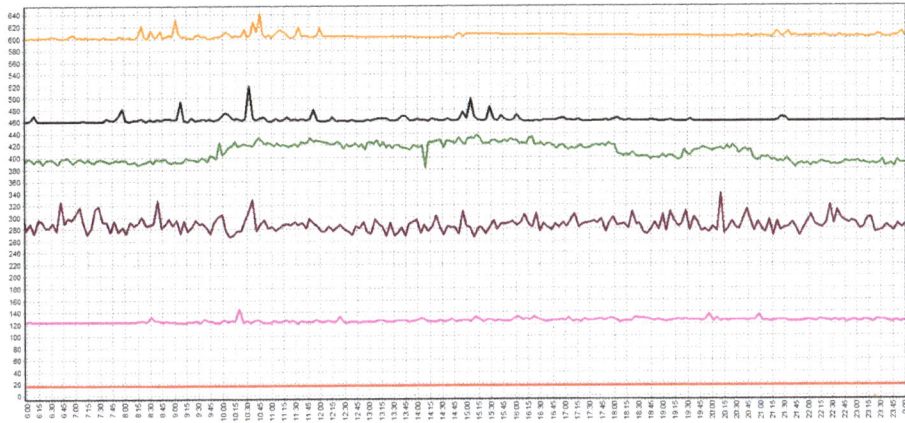

Figure 10. Siazan, Qum Island, Cybernetic, Neftchala, Shirvan, Turkmen01: 2014-06-06 Caspian Sea, Offshore Turkmenistan.

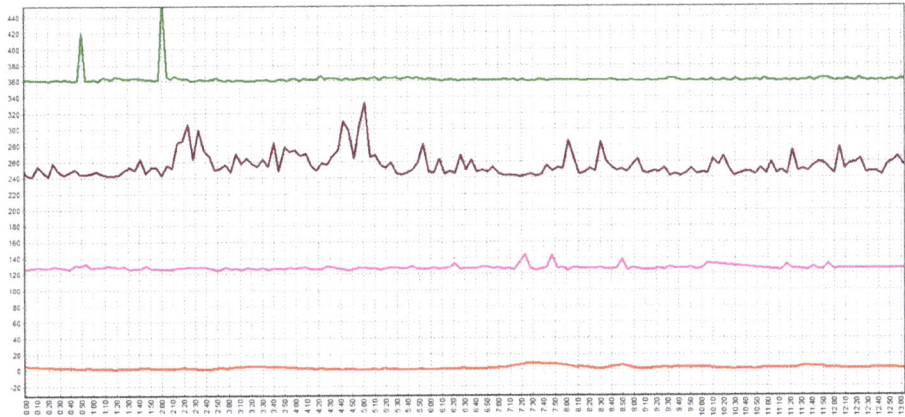

Figure 11. Siazan, Qum Island, Shirvan, Qazakh: 2014-06-29 Azerbaijan.

Row 3 presents us the results of the identification of location for the earthquake that took place on 16 September 2013 in the south of the Russian Federation.

From the charts of the third and fourth earthquakes (**Figure 4**), we can see that the ASP came from the south-east of the Caucasus and were recorded by the stations in the following order: Siazan—05:30, Qum Island—08:00, Qazakh—10:00, Neftchala—14:30. The system concluded that such a sequence corresponds to earthquake area VII, which is the north-east of Azerbaijan, where an earthquake did occur at 16:00/17:00 Baku time. The area was located at nearly 15 h before the actual earthquake.

Figure 5 shows that the system identified the coming earthquake area on the Iran-Iraq border from the combination of the times of ASP registration by the stations (Nakhchivan—08:00; Qum Island—09:00; Neftchala—08:50) 12 h before the earthquake.

Figure 6 relates to the earthquake that occurred at approximately 12:00 on 9 January 2014 in the Caspian Sea, Offshore Azerbaijan, and was registered 16 h before it occurred: by the Turkmen01 station at 09:15, by the Qum Island station at 09:25, by the Siazan station at 09:45, by the Neftchala stations at 11:15.

Row 6 of the table shows that the system identified the area of the expected earthquake in Turkmenistan. According to the chart in **Figure 7**, based on its sequence of registration by the stations (09:30 by Neftchala, 10:45 by Siazan, 11:30 by Qum Island), the system located the coming earthquake in Turkmenistan, i.e. in area I. The identification of the earthquake area took place over 24 h before the earthquake itself was registered.

Row 7 of the table contains the data related to the identification of location for the earthquake that occurred on 28 January 2014 in the western regions of Iran. According to **Figure 8**, based on the order, in which it was registered by the stations (Qum Island—09:45, Shirvan—07:30, Nakhchivan—04:50, and Neftchala—11:20), the system identified the location as area IX (Western Iran).

The data in Row 8 is related to the results of the identification of the area of the earthquake that occurred on 10 February 2014 in Azerbaijan. The curves of the chart (**Figure 9**) indicate that the sequence, in which the stations registered the ASP of that earthquake, was as follows: Qum Island—17:45, Shirvan—12:45, Qazakh—19:00 and Nakhchivan—18:30. Consequently, the system identified the area of the anticipated earthquake as area IV 19 h before the earthquake occurred.

Row 9 of **Table 2** contains the results of identification of the area for the earthquake that occurred on 7 June 2014 in Offshore Turkmenistan. The ASP of that earthquake was first registered by the Neftchala station at 06:45, then by the Qum Island station at 07:55, by the Siazan station at 08:15, by the Cybernetic station at 09:55, and finally by the Shirvan station at 10:20 (**Figure 10**).

The data in Row 10 is the results of the identification of the location of the anticipated earthquake that occurred on 29 June 2014 in Azerbaijan. **Figure 11** demonstrates that the order, in which an anomaly was registered by the stations, was as follows: Siazan—00:50, Qum Island—02:15, Shirvan—07:20, Qazakh—07:30. The system located the anticipated earthquake in area I (Azerbaijan).

7. Conclusions

1. The intelligent system based on the network of the RNM ASP stations and an ES combined with a neural network system can be used in locating the area of an anticipated earthquake. With the information on the direction and the number of the area of the anticipated earthquake, current combinations of the ASP, as well as the amount, list, date and magnitude of similar combinations registered in that area during previous earthquakes, a seismologist can assess the adequacy of the information on the location of the area of the anticipated earthquake. Having sufficient amount of time before the actual earth-

quake, the seismologist can bring in other specialists to participate in the decision-making to exclude a chance of error.

2. The stations comprising the RNM ASP network in our system are installed on wells of varying depths. Therefore, they have different characteristics, which are not easy to account for while locating the area of an anticipating earthquake and finding its magnitude.

    Besides, the deeper the well, the more expensive it is, which complicates the construction and maintenance of RNM ASP stations in the countries with no suspended oil wells available.

    In view of the above, our recommendation for the future is to build a network of stations standing on 50–100 m deep water wells, in which hydrophones would be placed in the water column at a depth of 10–20 m. We have established experimentally that more trustworthy results would be obtained by a network consisting of a large number (more than 10–15) of stations installed on wells of equal depth and located at equal distance from one another. Expanding the RNM ASP network to the countries in several seismically active regions via satellite communication can, in the long term, give a significant improvement in the results of determining the coordinates of the location of an anticipated earthquake.

3. We have also established experimentally that the efficiency of ASP monitoring and identification of the location of the area of an anticipated earthquake is directly proportional to earthquake strength. With the earthquake intensity exceeding five points, the identification of the earthquake location almost always gives valid results. The value of the estimate of the cross-correlation function $R_{X\varepsilon}(\mu)$ between the useful signal $X(i\Delta t)$ and the noise $\varepsilon(i\Delta t)$ decreases as the distance from the earthquake area grows. The value of the estimate of noise variance D_ε increases with the distance from the area; the correlation $R_{X\varepsilon}(\mu)/R_{X\varepsilon\varepsilon}(\mu)$ decreases with distance and $D_\varepsilon/R_{X\varepsilon\varepsilon}$ increases. The speed, with which the seismic-acoustic noise spreads in different types of medium, e.g. water, sand or clay, substantially varies. The well depth and the radius of the ASP monitoring correlate.

4. According to the results of the experiments on the Qum Island station, the range of that station significantly exceeds that of the stations standing farther from the Caspian Sea. The Siazan and Neftchala stations are located near the Caspian Sea and also have a wider monitoring range compared with other stations. Practically all seismic processes reaching the Caspian Sea are distinctly registered by them. The conclusion is that when building networks of new stations, we must account for the fact that the sea is a perfect conductor for seismic-acoustic noises that appear during incipient ASPs in the region.

5. Following the experimental data, we can suggest that the lead time of the registration of ASP origin by a seismic-acoustic RNM ASP stations over standard seismic equipment is determined by two factors.

    The first factor is that seismic-acoustic waves appearing at the start of an incipient ASP cannot reach the Earth's surface due to the frequency characteristics of some upper layers and spread instead horizontally as noise in deeper layers. When seismic waves reach the steel pipes of a well, they turn into acoustic signals and ascend at the velocity of sound to the surface to be caught by a hydrophone. Low-frequency seismic waves of seismic processes are registered by the receiving equipment of regular above-ground stations much later. By that time, an earthquake is already in progress.

The second factor is that with the use of noise technologies by analysing seismic acoustic noise, we can register incipient ASPs right at the start, as a correlation appears between the useful signal and the noise.

These two factors explain how RNM ASP stations are capable of indicating the time of the start of ASP origin so much earlier than the seismic survey service's above-ground stations.

6. Stations monitoring ASP can also monitor the hidden period of volcano formation well before a volcano erupts, or be used (on a regional scale) for testing minor and major nuclear bombs, as well as in assisting with other experiments related to the manufacture of military equipment.

Author details

Telman Aliev

Address all correspondence to: director@cyber.az

Institute of Control Systems of the Azerbaijan National Academy of Sciences, Baku, Azerbaijan

References

[1] Kanamori H, Brodsky EE. The physics of earthquakes. Reports on Progress in Physics. 2004;67:1429–1496. DOI: 10.1088/0034-4885/67/8/R03

[2] Ghahari F, Jahankhah H, Ghannad MA. Study on elastic response of structures to near-fault ground motions through record decomposition. Soil Dynamics and Earthquake Engineering. 2010;30(7):536–546. DOI: 10.1016/j.soildyn.2010.01.009

[3] Aliev TA, Abbasov AM, Aliev ER, Guluyev GA. Digital technology and systems for generating and analyzing information from deep strata of the Earth for the purpose of interference monitoring of the technical state of major structures. Automatic Control and Computer Sciences. 2007;41(2):59–67.

[4] Aliev TA, Guluyev GA, Pashayev FH, Sadygov AB. Noise monitoring technology for objects in transition to the emergency state. Mechanical Systems and Signal Processing. 2012;27:755–762. DOI: 10.1016/j.ymssp.2011.09.005

[5] Esref Y, Serhat T, Ali P. Analysis of ambient noise in Yalova, Turkey: discrimination between artificial and natural excitations. Journal of Seismology. 2013;17(3):1021–1039. DOI: 10.1007/s10950-013-9370-7

[6] Tomas F, Martin B. Detection capability of seismic network based on noise analysis and magnitude of completeness. Journal of Seismology. 2014;18(1):137–150. DOI: 10.1007/s10950-013-9407-y

[7] Kislov KV, Gravirov VV. Earthquake arrival identification in a record with technogenic noise. Seismic Instruments. 2011;47(1):66–79. DOI: 10.3103/S0747923911010129

[8] Yee E, Stewart JP, Schoenberg FP. Characterization and utilization of noisy displacement signals from simple shear device using linear and kernel regression methods. Soil Dynamics and Earthquake Engineering. 2011;31(1):25–32. DOI: 10.1016/j.soildyn.2010.07.011

[9] Aliev TA, Abbasov AM, Guluyev GA, Pashayev FH, Sattarova UE. System of robust noise monitoring of anomalous seismic processes. Soil Dynamics and Earthquake Engineering. 2013;53:11–25. DOI: 10.1016/j.soildyn.2012.12.013

[10] Aliev TA, Abbasov AM, Guluyev GA, Pashayev FH, Sattarova UE. Technologies and systems for minimization of damage from destructive earthquakes. In: Etirmishli G, editor. Seismoforecasting Researches Carried out in the Azerbaijan Territory. 1st ed. Baku: Nafta-Press; 2012. pp. 449–464.

[11] Hashemi M, Alesheikh AA. A GIS-based earthquake damage assessment and settlement methodology. Soil Dynamics and Earthquake Engineering. 2011;31(11):1607–1617. DOI: 10.1016/j.soildyn.2011.07.003

[12] Kanamori H. Real-time seismology and earthquake damage mitigation. Annual Review of Earth and Planetary Sciences. 2005;33(1):195–214. DOI: 10.1146/annurev.earth.33.092203.122626

[13] Colak OH, Destici TC, Ozen S, Arman H, Cerezci O. Frequency-energy characteristics of local earthquakes using discrete wavelet transform (DWT). World Academy of Science, Engineering and Technology. 2006;20:38–41.

[14] Hutton DV. Fundamentals of Finite Element Analysis. New York: The McGraw-Hill Companies; 2004. 494 p.

[15] Lockwood OG, Kanamori H. Wavelet analysis of the seismograms of the 2004 Sumatra–Andaman earthquake and its application to tsunami early warning. Geochemistry, Geophysics, Geosystems. 2006;7(9):Q09013. DOI: 10.1029/2006GC001272

[16] Aliev TA, Abbasov AM, Aliev ER, Guluyev GA. Patent "Method for monitoring and forecasting earthquakes". 2006 (International Application No PCT/AZ2006/00000, Pub.No WO2007/143799, International Filing Date June 16, 2006)

[17] Descherevsky AV, Lukk AA, Sidorin AY, Vstovsky GV, Timashev SF. Flicker-noise spectroscopy in earthquake prediction research. Natural Hazards and Earth System Sciences. 2003;3:159–164. DOI: 10.5194/nhess-3-159-2003

[18] Aliev TA. Digital Noise Monitoring of Defect Origin. New York: Springer; 2007. 224 p. DOI: 10.1007/978-0-387-71754-8

[19] Shebalin P, Keilis-Borok P, Gabrielov A, Zaliapin I, Turcotte D. Short-term earthquake prediction by reverse analysis of lithosphere dynamics. Tectonophysics. 2006;413(1–2): 63–75. DOI: 10.1016/j.tecto.2005.10.033

[20] Papagiannopoulos GA, Beskos DE. On a modal damping identification model for building structures. Archive of Applied Mechanics. 2006;76(7):443–463. DOI: 10.1007/s00419-006-0046-4

[21] Papagiannopoulos GA, Beskos DE. On a modal damping identification model for non-classically damped structures subjected to earthquakes. Soil Dynamics and Earthquake Engineering. 2009;29(7):29:583–589. DOI: 10.1007/s00419-006-0046-4

[22] Zafarani H, Noorzad A, Ansari A, Bargi K. Stochastic modeling of Iranian earthquakes and estimation of ground motion for future earthquakes in Greater Tehran. Soil Dynamics and Earthquake Engineering. 2009;29(4):722–741. DOI:10.1016/j.soildyn.2008.08.002

[23] Alcik H, Ozel O, Wu YM, Ozel NM, Erdik M. An alternative approach for the Istanbul earthquake early warning system. Soil Dynamics and Earthquake Engineering. 2011;31(2):31:181–187. DOI: 10.1016/j.soildyn.2010.03.007

[24] Rydelek P, Pujol J. Real-time seismic warning with a two-station subarray. Bulletin of the Seismological Society of America. 2004;94(4):1546–1550. DOI: 10.1785/012003197

[25] Satriano C, Wub Y-M, Zollo A, Kanamori H. Earthquake early warning: concepts, methods and physical grounds. Soil Dynamics and Earthquake Engineering. 2011;31(2):106–118. DOI: 10.1016/j.soildyn.2010.07.007

[26] Stankiewicz J, Bindi D, Oth A, Parolai S. Designing efficient earthquake early warning systems: case study of Almaty, Kazakhstan. Journal of Seismology. 2013;17(4):1125–1137. DOI: 10.1007/s10950-013-9381-4

[27] Tsuboi S, Saito M, Kikuchi M. Real-time earthquake warning by using broadband P waveform. Geophysical Research Letters. 2002;29(24):2187–2191. DOI: 10.1029/2002 GL016101

[28] Aliev TA, Abbasov AM, Mamedova GG, Guluyev GA, Pashayev FG. Technologies for noise monitoring of abnormal seismic processes. Seismic Instruments. 2013;49(1):64–80. DOI: 10.3103/S0747923913010015

[29] Aliev TA, Alizadeh AA, Etirmishli GD, Guluyev GA, Pashayev FG, Rzaev AG. Intelligent seismoacoustic system for monitoring the beginning of anomalous seismic process. Seismic Instruments. 2011;47(1):15–23. DOI: 10.3103/S0747923911010026

[30] Pujol J. Earthquake location tutorial: a graphical approach and approximate epicentral locations techniques. Seismological Research Letters. 2004;75(1):63–74. DOI: 10.1785/gssrl.75.1.63

[31] Sambridge M, Ghallagher K. Earthquake hypocenter location using genetic algorithms. Bulletin of the Seismological Society of America. 83(5):1467–1491.

[32] Khashei M, Bijari M. An artificial neural network (p, d, q) model for time series forecasting. Expert Systems with Applications. 2010;37(1):479–489.

[33] Rojas R. Neural Networks. 1st ed. Berlin: Springer-Verlag; 1996. 502 p. DOI: 10.1007/978-3-642-61068-4

2

Issues of the Seismic Safety of Nuclear Power Plants

Tamás János Katona

Additional information is available at the end of the chapter

Abstract

Seismic safety of nuclear power plants became an eminent importance after the Great Tohoku earthquake on 11th of March, 2011 and subsequent disaster of the Fukushima Dai-ichi nuclear power plant. Intensive works are in progress all over the world that include review of the site seismic hazard assessment, revision of the design bases, evaluation of vulnerability, and development of accident management capabilities of the plants. The lessons learned from the Fukushima-accident changed the paradigm of the design. Preparedness to the impossible, i.e. the development of means and procedures for ensuring the plant safety in extreme improbable situations became great importance. Main objective of the Chapter is to provide brief insight into the actual issues of seismic safety of nuclear power plants, provide interpretation of these issues, and show the possible solutions and scientific challenges. The "specific-to-nuclear" aspects of the characterisation of seismic hazard, including fault displacement are discussed. The actual design requirements, safety analysis procedures are briefly presented with main focus on the design extension situations. Operation aspects and problems for restart after earthquake are also discussed. The Chapter is more focusing on seismic safety of the inland plants, located on soil sites, in low-to-moderate (diffuse) seismicity regions.

Keywords: design basis, ground motion, ground displacement, defence-in-depth, design extension, liquefaction, safety analysis, margins, operation

1. Introduction

The nuclear catastrophe at the Fukushima-Dai-ichi plant caused by the Great Tōhoku earthquake followed by a huge tsunami on 11th of March 2011 alarmed worldwide attention to the safety of nuclear power plants (NPP). Enormous natural effects caused the accident.

However, the devastating consequences would have been limited or even avoided, if the provisions for tsunamis in the original design would be adequate. The nuclear catastrophe triggered actions worldwide: a comprehensive, complementary safety reviews, i.e. "stress-tests" have been launched in European Union member states just after 11th of March 2011. Similar programmes have been implemented in all countries operating nuclear power plants. The stress-tests have been aimed to the review of seismic hazard assessments for sites of nuclear power plants and to the verification of the design bases, as well as to the evaluation of margins against earthquake effects. Prompt and long-term measures have been decided by the nuclear operators for improvement of the accident management capabilities of the plants. The case of Fukushima Dai-ichi plant shows that the proper definition of the design basis hazard effects has to include also the phenomena generated by the earthquake (e.g. tsunami, soil liquefaction) and thorough checking whether the beyond design basis hazard effects can cause cliff-edge effect, i.e. sudden loss of safety functions due to effects exceeding the design basis one. Updated information for these programmes is provided at http://www.ensreg. eu/EU-Stress-Tests for the European Union, for the United States at http://www.nrc.gov/ reactors/operating/ops-experience/japan-dashboard.html and for Japan at http://www.meti. go.jp/english/earthquake/nuclear/.

The lessons learned gave essential feedback for upgrading the safety of operating reactors (see Refs. [1, 2]) and challenged the philosophy of the design of new plants. Instead of the "design for sufficient low probability of effects for ensuring the acceptable risk", the new design paradigm is "to be prepared for the impossible". Since an accident can never be completely ruled out, the necessary provisions for dealing with and managing a radiological emergency situation, onsite and offsite, must be planned, tested and regularly reviewed [3, 4].

Despite the severe accident of the Fukushima Dai-ichi plant, the nuclear power plants survive earthquakes. The International Atomic Energy Agency International Seismic Safety Centre collected the information on the earthquake experiences reported by the operators. More than two hundred magnitude >6 earthquakes have been registered (mainly in Japan) within 300 km epicentral distance from nuclear power plants (NPP) [5]. In most of the cases, there is no damage reported manly because of negligible effects at the site. However, there are important cases, including the Tōhoku earthquake, when the consequences have been either serious or enlightening for upgrading the operating and design of new nuclear power plants. Brief summary of experiences is given in Ref. [6]. One of the first event happened in Armenia when the Medzamor Nuclear Power Plant experienced minor shaking by Spitak earthquake (magnitude 6.8) in 1988. The epicentre was about 75 km from the plant. Although no damage occurred there, the plant was closed for 6 years due to safety concerns.

The most important lessons learned can be extracted from the earthquake experiences of nuclear power plants in Japan, where after singular warning cases the 11 March 2011 Tōhoku earthquake brought to temporary phaseout all nuclear power plants for radical safety review and upgrade.

The consequences of the Miyagi earthquake (August 2005, magnitude 7.2) at the Onagawa NPP were negligible, although the recorded ground motions exceeded those the plant was designed for. The plant was restarted 5 months after the earthquake. The Onagawa Nuclear

Power Plant was the closest nuclear power plant to the epicentre of the 11 March 2011 Tōhoku earthquake. Contrary to the Fukushima Dai-ichi plant, the 14-m-high seawall protected the Onagawa NPP from flooding. All safety systems functioned as designed, the reactors automatically were shutdown, and no damage of safety-related systems, structures and components (SSCs) occurred. A fire broke out in the turbine hall that did not challenge the plant safety [7]. The case of Onagawa NPP demonstrates that the proper definition of the design basis, e.g. the tsunami height is essential precondition of the safety. The Onagawa NPP has successfully passed the stress-test launched after Fukushima Dai-ichi accident in Japan [8]. The plant is ready to restart.

In July 2007, the Chūetsu offshore earthquake (Mw = 6.6) hit the Kashiwazaki-Kariwa NPP, the largest plant in the world. The experienced ground motions exceeded significantly the design basis level (0.69 g compared to the safe shutdown level of 0.45 g). Although there was no damage to the safety systems, the thorough proof of the plant post-earthquake condition and the reassessment of the seismic safety, which includes also re-evaluation of the site seismicity, identified the needs of certain upgrading measures, e.g. establishment of plant own fire-fighting capability [9]. The plant was idle for 21 months after the earthquake than the units No 1 and 5–7 have been restarted and operated up to April 2011. The units 2–4 have been not restarted. After 2011 Tōhoku earthquake, the plant was shutdown, stress-test and safety improvements have been carried out. The plant is not in operation.

In spite of the severe accident of the Fukushima Daiichi plant caused by the tsunami after Great Tohoku earthquake, the behaviour of 13 nuclear units in the impacted area on the East-shore of the Honshu Island demonstrated high resistance against ground vibrations due to earthquake. It seems, that the design of nuclear power plants complying with state-of-the art nuclear safety regulations and acceptable in the nuclear industry codes and standards ensure sufficient capacity to withstand the ground vibratory effects of earthquakes. The stress-tests of reactors in Japan after 11 of March 2011 resulted in justification of the external hazard design basis, quantification of margins beyond design basis and enhancing the tsunami protection of the sites, as well as in the improving the severe accident management capabilities of the plants. The capability of plants to withstand beyond design basis vibratory motion has been found to be adequate. For example, the stress-test of the Ohi NPP found that the units 3 and 4 would be able to withstand an earthquake with ground acceleration of up to 1.260 g that is exceeding 1.8 times its design basis of 0.7 g [10].

The stress-test in Japan also resulted in the strengthening of the regulatory structure and safety requirements. In October 2012, the new, independent from the industry Nuclear Regulation Authority (NRA) has been established. NRA announced that henceforth nuclear power plant restart reviews would comprise safety assessment by NRA based on safety guidelines in the new regulatory requirements [11]. Main focuses of the reformed safety requirements are as follows: emphasis on defence-in-depth concept, assessment and enhancement of the protective measures against extreme natural hazards, measures against severe accidents (and terrorism), elimination of common cause failures, back-fitting to the existing plants. One of the new focuses of the NRA was the proofing of the potentially active faults at the nuclear sites and development standards concerning displacement and ground deformation in addition to those for seismic ground motion. The new requirements with regard to the existence of active faults at the plant vicinity have been enforced in July 2013, and a methodical study has been published in September 2013 [12]. The active faults at the site vicinity affect the design basis ground motion and depending on the distance from the plant they can cause permanent disruption of the ground that can impact the safety functions of the SSCs. The faults that can cause permanent surface disruption are generally called capable, exceptions are Japan and Russia, where the term active fault is still used for structures capable to cause surface deformation. In earlier Japanese practice, the nearby active faults have been integrated into the deterministic hazard assessment that resulted in very high design basis accelerations. Especially after revision of the regulation in 2006, when the magnitude of the just below the site source has been increased from the value of 6.5–6.7. The earthquake experience demonstrated that the vibratory motion can be managed by proper design, but the surface rupture below or in the very vicinity of the safety-related structures can cause very significant and not properly studied effects, e.g. relative movements below the foundations, tilting of the structures. The issue has been already recognized before the accident of the Fukushima Dai-ichi NPP; see Ref. [13]. By definition, the fault is considered active if it shows evidence of past movements a recurring nature within such a period that it is reasonable to conclude that further movements at or near the surface may occur. The time frame has been increased from 50,000 years to 125,000 years (in 2006) and now the Japanese Nuclear Regulatory Authority requires 400,000 years in uncertain cases. In 2012–2013, it was recognized that the shatter zones beneath the Higashidori NPP are likely to be active, seismogenic faults. The fractures at the Tsuruga NPP, that lie close to or pass beneath Unit 2, could also be active faults. Clarification of the nature of below site faults became condition of the restart for Higashidori and Tsuruga NPPs, thus the operators invited independent investigation teams to review and to ascertain the scientific validity of the fault activity [14, 15]. In 2016, the Nuclear Regulatory Authority indicated that at four other NPP sites (Ohi, Mihama, Shika and Monju) might also have active faults [16–19].

As of April 2016, total of 16 pressurized water reactors (PWRs) and 10 boiling water reactors (BWRs) have filed application for the conformity review at the Nuclear Safety Authority. Out of 16 PWR cases, five PWRs received the NRA's permission for changes in reactor installation and two PWRs have been restarted [20].

A conscious development of the seismic safety of nuclear power plants can be recognized in the United States. In the past several decades, the Nuclear Regulatory Commission (NRC) and the industry have undertaken a number of initiatives to address potential plant vulnerabilities to natural phenomena. In 1977, the NRC initiated the systematic evaluation program to review among the others seismic designs of older operating nuclear reactor plants in order to reconfirm and document their safety. In 1980, the NRC established the Unresolved Safety Issue A-46 program that was focused on the seismic re-qualification of some mechanical and electrical equipment of the operation plants for the design basis ground motion. In 1991, the NRC initialized the program for Individual Plant Examination of External Events (IPEEE) for severe accident vulnerabilities that included evaluation of seismic safety and verification of the seismic adequacy of equipment for the design basis earthquake; see Ref. [21]. Following the processes in the United States, the OECD countries (except Japan) also performed some limited re-evaluation of their older nuclear power plants for seismic events [22]. As it is mentioned above, Japan started to consider the seismic safety re-evaluation of NPPs after Kashiwazaki-Kariwa event. Most extensive programmes have been performed in Eastern European countries. These programmes were motivated mainly by the strengthening of the regulations, changing the understanding of the site seismic hazard and consequently, establishing new seismic design basis. The operators of WWER reactors implemented comprehensive programmes for evaluating and upgrading the seismic safety of their nuclear power plants [23–25].

In the past decades, thanks to the abovementioned seismic safety programmes, essential methodological developments have been performed in the regulation as well as in the practice of seismic hazard assessment, design and qualification procedures, quantification of margin with respect to the design basis and seismic safety analysis. Probabilistic seismic hazard assessment (PSHA) methodology has been developed and adopted for the re-assessment and updates of the seismic hazard characterization. Contrary to the deterministic seismic hazard assessment, the PSHA accounts the randomness of the natural phenomena as well as the epistemic uncertainties. The hazard curve from PSHA is also to use in the seismic probabilistic safety analyses that provide quantitative judgement of safety with respect to earthquakes. The methodological developments have been made mainly in the United States and have been promoted also by International Atomic Energy Agency (IAEA) guidelines. The divergence between the practice of Japan and the countries following the IAEA guidelines has become obvious after Chūetsu offshore earthquake while the situation Kashiwazaki-Kariwa NPP have been assessed by the international community [9, 26]. The United States approach regarding seismic margin evaluation has been justified in the practice as 23 August 2011, a 5.8 magnitude earthquake occurred, 11 miles from the North Anna NPP, United States, Virginia. Although the ground motion experienced at the site exceeded the design basis, the plant survived the earthquakes without significant damages as it was to expect on the basis of seismic margin assessment [27].

The advances of the United States regulation and practice became more explicit after the severe accident at Fukushima Dai-ichi NPP [28]. The basic statement and starting point of the post-Fukushima actions in the United States was that "the continued operation and licensing of nuclear power plants do not pose an imminent risk to safety" [29]. This confident but critical approach to the Fukushima-issue resulted in a very systematic evaluation of lessons learned and rational definition of the actions in the United States that is essentially differing to the reaction of many professionals and especially the officials in Japan and in European Union. The way how to implement the seismic near-term task force recommendations of the NRC in the area of seismic safety are defined in Ref. [30]. The screening is based on the development of the site-specific ground motion response spectra (GMRS) in accordance with the Regulatory Guide 1.208 [31] and their comparison to the design basis safe shutdown earthquake (SSE) response spectra. For the plants where the GMRS exceeds the SSE additional actions have to be implemented for justification of seismic safety and margin to withstand the beyond design basis earthquakes. An Expedited Seismic Evaluation Procedure (ESEP) was developed to focus initial resources on the review of a subset of the plant equipment that can be relied upon to protect the reactor core following beyond design basis seismic events [32].

In the European Union, the focused safety assessment of the nuclear power plants did not reveal dramatic safety deficiencies [33]. In general, the seismic design basis is satisfactorily determined on the basis of events consistent with a 10^{-4} per annum return frequency. The existence of necessary margins with respect to the earthquakes exceeding the design basis one that ensures to avoid the cliff-edge effects. In some countries, the design basis horizontal peak ground acceleration (PGA) has to be set for 0.1 g in compliance with IAEA guidance [34, 35]. The active/capable fault issue has also been addressed at some plants; see for example Slovenian stress-test report in Ref. [36]. The stress-tests did not identify dramatic safety deficiencies at European NPPs. In all countries, the stress-test resulted in the strengthening of the safety requirements, consequent implementation of the defence-in-depth concept, an improvement of severe accident management and mitigation of severe accident consequences accounting for the specific issues at multi-unit sites. The whole stress-test process resulted in strengthening the regulatory requirements and harmonization of national regulations; see Refs. [37, 38].

The recent status of seismic safety of nuclear power plants demonstrate that the acceptable level of safety can be assured, and the robustness of the plants can be demonstrated, if the definition of the seismic hazard is adequate and updated regularly. Nevertheless, the earthquakes might be the dominating contributors to the overall risk of nuclear power plants as it has been demonstrated by the seismic probabilistic safety assessments (SPSA) of several nuclear power plants. It is mainly because of large uncertainty of the characterization of seismic hazard and complex behaviour of the plant hence all systems, structures and components are affected by the earthquake.

The safety aspects of operation of nuclear power plants have primary importance. However, the operation of nuclear power plants cannot be ignored. The nuclear power production has enormous economic importance in many countries [39, 40]. Consequently, the issue of safe continuation or restart of operation after an earthquake has also a great importance. Obviously, there is a need for reliable justification of plant safe status after earthquake for avoiding long shutdown time and consequent economic losses. A rapid assessment of the post-event plant status is very important for assessing the conditions for restart or in extreme cases for assessing the plant condition for emergency management. With this respect, the case of Kashiwazaki-Kariwa NPP demonstrates the negative experience and the positive one is the case of North-Anna NPP. The lessons learned from the Fukushima Dai-ichi accident shows the importance of adequate judgement on the post-earthquake plant condition.

Main objective of the Chapter is to provide brief insight into the actual issues of seismic safety of nuclear power plants. The most important and "specific-to-nuclear" aspects of the characterization of earthquake and associated with earthquakes hazards and definition of the design basis ground motion and fault displacement are presented. The change by Fukushima accident paradigm of safety and its manifestation in design, safety evaluation and operation are in the focus of the chapter. The intention is to provide information on the seismic safety issues that is relevant mainly for the inland plants, located on soil sites, in low-to-moderate (diffuse) seismicity regions. The chapter structured as follows:

- Overview of the burning issues of the seismic safety of nuclear power plants in the introduction
- Overview of basic safety requirements
- Peculiarities of site characterization and definition of the design basis and design basis extension
- Basic design requirements focusing more on the design extension issues
- Brief overview of the seismic safety assessment issues
- Operational aspects of seismic safety, Earthquake preparedness, procedures
- Restart after earthquake, post-event inspections, damage indicators.

The chapter is not intended to provide textbook information either for the characterization of site seismic hazard or for the aseismic design of nuclear power plants. Definition of the generic terms of seismology and seismic engineering are assumed to be known. The terms related to seismic safety of the nuclear power plants are explained, only.

2. Basic principles of seismic safety

2.1. Safety objectives

The fundamental safety objective of design and operation of nuclear power plant is to protect people and the environment in case of any malfunctions, failures of the plant systems, structures and components, which may occur during the plant lifetime including those caused by rarely occurring earthquakes [41]. The earthquakes affect the site and plant in a complex manner, by vibratory ground motion, permanent ground deformations, and seismic induced ground failure or flooding. There are also other phenomena related to the tectonic environment that can affect the plant safety, e.g. tectonic creep, after-slip, uplift and subsidence. Earthquakes are the most challenging the safety external hazards that affect simultaneously all items important to safety, including systems to manage severe accidents. Widespread failures at the plant and site, and in surrounding area can hindrance to human intervention.

From technical point of view, to protect the human life and the environment in case of earthquakes, the fundamental safety functions have to be ensured that are as follows:

- Control of reactivity in the reactor and spent fuel pool;

- Removal of heat from the reactor and from the spent fuel pool;

- Confinement of radioactive material, shielding against radiation, as well as limitation of accidental radioactive releases.

Those systems, structures and components that are necessary to fulfil the fundamental safety functions and/or affect the fundamental safety functions are indicated as "items important to safety".

2.2. Compliance with safety objective

2.2.1. Concept of the design defence in depth

According to the Principle 8 in IAEA Safety Fundamentals [41], the primary means of preventing and mitigating the consequences of accidents is 'defence-in-depth'. Defence-in-depth is a systematic combination of consecutive and independent levels of protection. If one level of protection or barrier were to fail, the subsequent level or barrier would be available. The levels of defence are as follows:

1. Prevention.

2. Control of anticipated operational occurrences.

3. Control of accidents.

4. Mitigate the release in a severe accident via ensuring integrity and leak-tightness of the containment so as to prevent the exceedance of the severe accidents release limits (for example, protection the containment from the hydrogen explosion).

5. Mitigation of consequences of the radiological consequences to the population in the event with release of considerable amounts of radioactive substances by emergency preparedness arrangements.

The entire system for ensuring the plant safety is given in **Table 1**; see Ref. [3].

Levels		Objective	Essential means	Radiological consequences	Associated plant condition categories
Level 1		Prevention of abnormal operation and failures	Conservative design and high quality in construction and operation, control of main plant parameters inside defined limits	Operation within the authorized limits	Normal operation
Level 2		Control of abnormal operation and failures	Control and limiting systems and other surveillance features		Anticipated operational occurrences
Level 3	3.a	Control of accident to limit radiological releases and prevent escalation to core melt conditions	Reactor protection system, safety systems, accident procedures	No off-site radiological impact or only minor radiological impact	Postulated single initiating events
	3.b		Additional safety features, accident procedures		Design Extension Conditions: Postulated multiple

Levels	Objective	Essential means	Radiological consequences	Associated plant condition categories
				failure events; Accident caused by a rare external events without severe fuel failure
Level 4	Control of accidents with core melt to limit off-site releases	Complementary safety features to mitigate core melt, Management of accidents with core melt (severe accidents)	Off-site radiological impact may imply limited protective measures in area and time	Postulated core melt accidents. Confined fuel melt – also considered as design extension condition
Level 5	Mitigation of consequences of significant releases of radioactive material	Off-site emergency response Intervention levels	Off-site radiological impact necessitating protective measures	–

Table 1. The structure of the levels of defence-in-depth.

The levels of defence belong to certain plant conditions that should have different annual probabilities, see **Table 2**.

Level	Denomination	Frequency (1/year)
Level 1	Normal operation	Quasi continuous
Level 2	Operational events	$f > 10^{-2}$
Level 3	Design basis accidents	$10^{-2} > f > 10^{-4}$
	Design extension accidents	$10^{-4} > f > 10^{-6}$
Level 4	Design extension accidents with limited but confined core melt	$10^{-6} > f$
Level 5	Severe accidents	

Table 2. Categorization of plant statuses according to annual frequency.

For sake of clarity, a brief definition of most important terms has to be performed here:

- **Design basis** refers to the range of conditions and events (i.e. earthquakes and phenomena associated with) taken explicitly into account in the design of a facility, according to the established criteria, such that the facility can withstand them without exceeding authorized limits by the planned operation of safety systems.

- **Design basis conditions** mean the normal operation, anticipated operational occurrences and accidents in which the degradation of the reactor core is excluded due to design features.

- **Design basis earthquake** (DBE) or safe shutdown earthquake (SSE) refers to the earthquakes affecting the plant site effect of those are used in the design of SSCs ensuring the basic safety functions as it is required by the nuclear safety regulations and in compliance with industrial standards. These effects are the vibratory ground motion and surface deformation. The design basis earthquake surface deformation is distortion of geologic strata at or near the ground surface by the processes of folding or faulting as a result of various earth forces. Tectonic surface deformation is associated with earthquake processes. The design basis earthquake has to be defined at required annual frequency of occurrence (return period) and confidence level. Design basis earthquake should not result in reactor core melt.

- The **operating basis earthquake** ground motion (OBE) is the vibratory ground motion for which al SSCs necessary for continued safe operation will remain functional. The operating basis earthquake ground motion is only associated with plant shutdown and inspection unless specifically selected by the plant owner as a design input.

- **Design extension conditions** refer to the accidents not considered in the design basis, including accidents with significant degradation of the reactor core. However, releases of radioactive material are kept within acceptable limits. The accidents caused by a rare external event, e.g. earthquakes exceeding design basis one, in which a considerable part of the fuel in a reactor or in a spent fuel pool loses its original structure are to classified for design extension. Some national regulations define two levels of design extension conditions:

  ◦ Design extension conditions 1 (DEC1) are consequences of complex sequences not accounted for in the design basis. The DEC1 should not result in core damage and the plant can be brought into safe shutdown condition.

  ◦ Design extension conditions 2 (DEC2) are severe accidents with core damage. However, the heat removal from the core can be established or restored in DEC 2 and the releases from the containment have to be limited.

- **Severe accidents** refer to the accidents in which considerable part of the fuel in reactor loses its integrity. Accident sequences with core melt resulting from external hazards which would lead to early or large releases should be practically eliminated. The practical elimination should be primarily based either on the obvious physically impossibility for the accidents to occur, or on the design provisions that eliminate these accident sequence with a high degree of confidence.

2.2.2. Acceptance criteria

The compliance with basic safety objectives needs proper qualitative measure. The "authorized limits", i.e. the dose restrictions guaranties the avoidance of adverse effects to the people and environment. From this point of view, the normal operation can be qualified to be acceptable if the annual dose of an individual in the population, arising from the normal operation of a nuclear power plant, is 0.1 mSv. The same limit is valid for anticipated operation occurrences. Limits for accident are usually set to 1 mSv/event or 5 mSv/event, depending on the probability of extension condition that can be defined by amount of radioactive substances released. For example, in Finland, in order to restrict long-term effects, the limit for the atmospheric release of cesium-137 is 100 TBq [42]. The possibility of exceeding the limit shall be extremely small. Even in the case of severe accidents, the release of radioactive substances shall not necessitate large-scale protective measures for the public nor any long-term restrictions on the use of extensive areas of land and water. The possibility of a release in the early stages of the accident requiring measures to protect the public shall be extremely small.

Probabilistic criteria for acceptance can be expressed in terms of annual frequency of core damage (CDF) or large early releases of radioactive substances (LERF) that are directly related to the design. Considering all the design basis operational statuses and initiating events collectively, the $CDF \leq 10^{-5}/a$ and $LERF \leq 10^{-6}/a$ are generally accepted events. In some countries, the limit 20 mSv is defined for design extension conditions. In some countries, the "limited environmental effect" is required in case of design.

2.3. How to ensure the seismic safety?

Traditionally, the design of the nuclear facilities adapted the two-level concept: design for safety, accounting for a high-level, low probability of exceedance seismic excitation for design basis and design for service, using a moderate level of seismic excitation for operational limit.

The design basis earthquake is denoted as safe shutdown earthquake (SSE) in accordance with the United States terminology; see Ref. [43]. It is called SL-2 earthquake level by the IAEA guideline NS-G-1.6 [35]. Here, the term of design base earthquake (DBE) will be used. According to the international practice, the annual probability of exceedance of the DBE is usually $10^{-4}/a$ in case of nuclear power plants. SSCs required for basic safety function have to sustain the earthquake loads without loss of function. In the traditional design philosophy, the plant condition after SSE corresponds to the ultimate limit states.

Operability of NPPs should be ensured after the more frequent and moderate severity earthquakes. The operational base earthquake (OBE or SL-1 level according to the IAEA terminology) level is defined as a design level for continuous operation [35]. The OBE was usually defined as an event with frequency of $10^{-2}/a$, or a ground motion with peak ground (horizontal) acceleration (PGA) that equals to a given fraction of PGA value of the SSE. Through the years, the concept of designing for two earthquakes has radically changed. Nowadays, the OBE is interpreted as an operational limit and inspection level rather than an obligatory design level. Setting the OBE level is matter of design, operational, economic considerations. For example, there is no need for specific design measures for an OBE, if its PGA is equal or less than 1/3rd of the SSE PGA. Generally, if the OBE level is exceeded, the automatic shutdown of the reactor is not required [43]. In the traditional design philosophy, the post-earthquake plant condition up-to OBE level corresponds to the serviceability limit state.

The application of the defence-in-depth concept (DiD) in the design modified the outlined above traditional design concept:

- Levels 1 and 2 ensure serviceability of SSCs of the entire plant, i.e. reliable operation up-to certain level of earthquake effects.

- Level 3.a ensures to the serviceability limit states of safety-related SSCs for design basis earthquake effects.

- Level 3.b ensures the irreversible serviceability limit states of safety-related SSCs for earthquake effects exceeding those accounted for in the design basis.

- Level 4 ensures to the ultimate limit state of safety-related SSCs for earthquake effects exceeding those accounted for in the design basis, and serviceability of SSCs that are dedicated for severe accident management (hardened core SSCs).

- Level 5 corresponds to the planning of disaster management using all SSCs that survived the earthquake including mobile, provisional and external equipment.

The defence-in-depth (DiD) hierarchy of protective means has to be in place for the case of earthquakes as it is shown in **Table 3**.

Level 1	In case of felt earthquake, the plant either continues to operate or shutdown automatically or manually. The criteria for safe continuation of operation have to be defined.
Level 2	If the operation is terminated but the safety systems have not been activated (except of the safe shutdown automation) means, plans and procedures have to be in place for assessing the post-earthquake conditions and restoring the operational conditions.
Level 3a	If the operation is terminated and the safety systems are activated, the plant has to be stabilized in safe condition and thorough inspection, re-assessment are needed for the decision of restoration or permanent shutdown of the reactor. The safety-related SSCs have to ensure the basic safety functions in case of design basis earthquakes.
Level 3b	The design extension condition with respect to earthquakes means that the systems, structures and components needed for fundamental safety functions shall have sufficient margin to withstand earthquakes effects exceeding those in the design basis and ensures the integrity of reactor core.
Level 4	The very rare earthquakes that are sufficient larger than the design basis one should not be excluded from the considerations. Early or large releases from the accident sequences with core melt should be practically eliminated. Effects of rare and severe earthquakes need to be considered in the design but realistic, best estimate methods and assumptions can be applied. In spite of the core damage, the radiological consequences have to be limited due to containment.
Level 5	Means, plans and procedures have to be in place for on-site and off-site emergency response to mitigate the consequences of accidents.

Table 3. Levels of defence-in-depth in case of earthquakes.

Generally, the seismic safety is ensured by the following complex activities:

1. Proper site selection

2. Site investigations and evaluation of the site seismic hazard, including associated with the earthquake hazards, e.g. tsunamis, surface rupture, soil liquefaction

3. Definition of the design basis earthquake and surface displacement

4. Adequate design with consequent implementation of the defence-in-depth concept and rules of aseismic design that includes:

    a. the suitable dimensioning and lay-out of structures that ensure the required seismic resistance and margins—avoiding cliff-edge effects;

    b. use of verified methodologies and standards applicable for nuclear industry;

    c. use of high-quality and qualified for vibratory effects products;

    d. use of appropriate instrumentation and protection systems;

5. Evaluation of safety—with feedback to the entire design process

6. Development of accident-prevention and accident-management procedures

7. Conscious operation, maintenance and seismic housekeeping

8. Periodic safety assessment, including re-evaluation of the hazard and subsequent upgrading if needed.

2.4. Site selection, site suitability

The desired seismic safety of NPPs can be ensured by proper selection of the site and adequate investigation and evaluation of site hazards.

The site selection is a preventive measure, while the site characterization is a mitigative measure that aims to limit the potential risk of the facility since it provides the basis for proper definition of the design basis.

It is obvious that the most rational way to protect the plant from the effects of hazards is to select a site with obvious low exposure.

The siting is a multiphase process [44] that starts with the survey of large area for potential sites. The area is defined on the basis of economic (e.g. area for economic development), technical (e.g. availability of cooling water) and safety considerations. The site investigation and hazard evaluation confirms the site selection and provides the necessary information for the derivation of the design basis. This confirmatory activity extends to the whole lifetime of the plant. It is now widely accepted that the hazard assessments have to be reviewed regularly that is an obligatory activity in majority of nuclear power plant operating countries. These periodic safety reviews provide the frame for integration the new scientific evidences and experiences into the site hazard characterization.

Nuclear power plants can be built practically anywhere. From safety point of view, there are a few criteria to be considered for site suitability. The sites shall only be qualified unsuitable, if it is concluded during characterization of external hazards that no engineering solutions exist to design protective measures against those hazards that challenge the safety of plant. The site suitability criteria are defined in the international regulatory documents (see generally in NS-R-3 in Refs. [44, 45] and particularly in national regulations, for example [46] for United States and [47, 48] for Russia.

There are also site features that affect the engineering effort needed for ensuring the plant safety. These features can be considered as discretionary criteria and can be used to facilitate the selection process. A global balance has to be established between the characteristics of a site, on the one hand, and specific design features, site protection measures in order to obtain the required level of safety.

The tectonic environment can affect the plant safety by ground motion due to earthquakes, permanent ground deformations, seismic induced ground failure, tectonic creep, after-slip, uplift and subsidence.

The intensity of ground motion (peak ground acceleration at the free-field) is usually not a matter of suitability considerations. This is subject of discretion that may affect the ranking and selection of the site. The only exception might be the national regulation of Russia. According to this, the sites have to be qualified as unacceptable where the intensity of maximum credible earthquake is $I \geq 9$ on MSK-64 [48].

From the point of view of tectonic environment and phenomena, it is unacceptable, if the reliable evidence shows the existence of a fault capable to cause permanent surface movement or dislocation that has the potential to affect the safety of the nuclear installation and cannot been compensated by engineering methods.

2.5. Site seismic hazard evaluation

The geological, seismological and geotechnical investigations performed for site evaluation needed to provide information to support the following:

1. seismic source characterization input to a probabilistic seismic hazard analysis (PSHA);

2. evaluation of surface fault rupture hazard;

3. site response analysis; and

4. evaluation of seismic-induced ground failure hazard.

The adequacy of the characterization of the hazards is a precondition of the adequacy of the design.

The site seismic hazard is characterized by

- hazard curve, i.e. function of annual probability of exceeding given value of peak ground acceleration,

- uniform hazard response spectra (UHRS)

at base-rock or outcrop and at the free-field.

For specific calculations, e.g. liquefaction analyses the deaggregation of the hazard, i.e. definition of main contributors to the hazard in form of magnitude-distance bins are needed.

For analyses in time domain, accelerograms are needed. These can be generated fitting to the UHRS or selected from earthquake records and tuned to the UHRS.

The rules and requirements for evaluation of site seismic hazard are given in the IAEA Safety Guide SSG-9 [34]. A brief overview of the subject is given below that focuses on specific "nuclear" aspects of the site seismic hazard evaluation.

3. Definition of the design basis

3.1. Probabilistic attributes of the seismic design basis

3.1.1. Design basis earthquake

The design basis earthquake has to be characterized by maximum ground motion acceleration defined at certain level of annual exceedance probability.

For the definition of exceedance probability compatible with the probabilistic safety targets (acceptable CDF and ELRF, see Section 2.1), let us assume that a single parameter can characterize the damage potential of the earthquake, e.g. the free-field maximum horizontal acceleration, PGA = a. The probabilistic criteria, $Pscr$, to be screened in to the design basis can be derived from the accepted probability (or annual frequency) of early large releases, PELR that is generally equal or higher than 10^{-7}/a (for plants of older design 10^{-6}/a), and the conditional probability of failure $P(ELR|PGA \geq a)$ due to a particular impact, the consequence of which is a release exceeding the regulatory limit.

Since the probabilistic target is $P_{scr} \times P(ELR|PGA \geq a) < P_{ELR}$, the screening probability can be written in a very simple way:

$$P_{scr} = \frac{P_{ELR}}{P(ELR|PGA \geq a)} \tag{1}$$

In this case, $Pscr$ means the probability of an impact affecting the facility, originating from the given source of hazard during, which causes damage that results in an early large release.

The $1/P(ELR|PGA \geq a)$ can be interpreted as the required performance of the SSCs, as it is given in the ASCE/SEI 43-05 [49], where the hazard exceedance probability, H_D and the target performance P_F is defined for SSCs with different importance for safety in a graded manner, i.e.

$$H_D = R_P \times P_F \tag{2}$$

where R_p indicates the probability ratio that practically identical with $1/P(ELR|PGA \geq a)$. For the SSCs in highest seismic design category, the $R_p = 10$ and $P_F = 10^{-5}/a$, consequently, the annual hazard exceedance probability is equal to 10^{-4}.

The uncertainty of definition of seismic hazard has to be accounted for in the design basis. Therefore, the confidence level of the hazard definition is also matter of nuclear safety regulation. Generally, the mean characteristics of the hazard are taken into account in the design basis. In some countries, for example, the $10^{-5}/a$ median value is accepted [50], while in other countries the 84 percentiles. It has to be emphasized, the probability distributions of peak ground acceleration at low exceedance levels are rather skewed, i.e. the mean value exceeds the median one, and will exceed the 84 percentiles at low exceedance levels.

3.1.2. Avoiding the cliff-edge effect: design extension conditions

It is important to avoid the cliff-edge effect, i.e. sudden degradation in case if the experienced earthquake effects exceed those accounted for in the design. It has to be emphasized that it is not a new, post-Fukushima invention, but it is part of safety philosophy since several decades.

For example, according to the ASCE/SEI 43-05 [49] and United States NRC Regulatory Guide 1.208 [51], the response spectra for design basis earthquake ground motion (DRS) has to be defined as $DRS = DF \cdot UHRS$, where DF is the design factor depending on the slope factor of the hazard curve, i.e. the ration of spectral amplitudes of the free-field UHRS calculated for the H_D and $0.1H_D$. If the hazard curve is steep, or otherwise, if the increase in spectral amplitude is moderate if the exceedance probability is decreasing by an order of magnitude, the $DF \approx 1$ and $DRS \approx UHRS$. In opposite case, the $DF > 1$ and $DRS > UHRS$. Thus, the above definition of design response spectra provides already the necessary assurance to avoid the cliff-edge effect.

3.1.3. Design basis of SSCs for severe accident management and mitigation

According to the defence-in-depth concept, the Level 3a corresponds to design basis conditions, and the Levels 3b and 4 correspond to the design extension conditions. The design basis of the SSCs having safety relevance in case of Levels 3a and 3.b can be defined as it is described above.

The irreversible serviceability in case of Level 3.b is ensured by designed-in margins.

It is also shown in the ASCE/SEI 43-05 that as long as the seismic demand and structural capacity evaluations have sufficient conservatism that grantee both less than about a 1% probability of unacceptable performance for the design basis ground motion, and less than about a 10% probability of unacceptable performance for a 1.5 times larger PGA than those for design basis. It is shown in the commentary of the code that the nominal factor of safety with respect to the design basis effects is not less than 1.5 with 10% conditional probability of failure. The hazard exceedance level corresponding to the 150% of design basis effect, i.e. to the design basis, depends on steepness of the hazard curve.

Consequently, the SSCs designed for $10^{-4}/a$ design basis earthquake in compliance with relevant standards will have sufficient margin for performing their intended safety function even if an earthquake with $PGA \leq 1.5 \times PGA$ at 10^{-4}/a. There are plant sites, for example, the Paks site in Hungary, where the hazard curve is rather steep and the 1.5 times of the PGA of the design basis earthquake corresponds to the exceedance probability more than one order of magnitude less than the design basis one [52].

In case of Level 4, the designed in margins of the safety relevant SSCs with passive function (e.g. containment) could ensure the integrity. However, the designated severe accident management SSCs have to survive the effects of very severe earthquake and have to be functional after the event. For the definition of the design basis of these SSCs-specific considerations are needed. This can be made, for example, on the basis of the "Eurocode: Basis of structural design".

Regarding the reliability of structures, the "Eurocode: Basis of structural design" Annex C [53] applies the first-order reliability method for demonstration of the reliability of the partial factor method of design. The probability of failure P_{fail} is defined by the reliability index, β

$$P_{fail} = \Phi(-\beta) \tag{3}$$

where Φ denotes the standardized normal distribution function. For the normally distributed random variable $g = R - E$, β is the ratio of mean value of g, and its standard deviation is, where R is the resistance and E is the effect of actions. For example, if target P_{fail} is equal to 10^{-6}, β = 4.75. The value of the reliability index depends on the importance of the structure, and on the limit state. The index β as well as the P_{fail} can be related to reference period 1 year or to the total service lifetime, e.g. 50 years.

As it is given in Section 2, the design basis earthquake causes an irreversible serviceability limit state, while a beyond design basis earthquake causes design extension condition, i.e. ultimate limit state. Thus, the value of β for the ultimate state and highest reliability class is equal to 5.2 for 1 year, and 4.3 for 50 years.

The failure boundary is, when $g = R - E = 0$. Generally, the design effect of action depends on the actions, geometrical properties of structure and model uncertainties, while the design resistance depends on the material properties, geometrical properties and model uncertainties. The design value of action effects E_d and the resistance R_d should be defined such that the probability having a more unfavourable value will be

$$P(E > E_d) = \Phi(+\alpha_E \beta) \tag{4}$$

$$P(R \leq R_d) = \Phi(-\alpha_R \beta) \tag{5}$$

where $|\alpha_E| \leq 1$ and $|\alpha_R| \leq 1$ are the sensitivity factors for effects of actions and resistance, respectively, and β is the target reliability index.

Let us consider the SCCs for severe accident management and mitigation and assume that these are of highest reliability class and assume that these SSCs have to be function at 10^{-5}/a mean hazard level and have to be in ultimate condition at 10^{-6}/a hazard level [52].

According to the Code, the minimum value of reliability index for the ultimate limit state and for 1-year reference period is $\beta = 5.2$, that corresponds to the target probability of failure $P_{fail} = 10^{-7}$/a, and $\beta = 4.75$ for the target probability of failure $P_{fail} = 10^{-6}$/a. According to the Code, the values $\alpha_E = 0.7$ and $\alpha_R = 0.8$ can be assumed for the sensitivity factors. Results of calculation of annual exceedance probabilities for effects and resistance are given in the **Table 4**; see Ref. [52].

	E		R	
P_{fail}	10^{-6}/a	10^{-7}/a	10^{-6}/a	10^{-7}/a
β	4.75	5.2	4.75	5.2
α	$\alpha_E = -0.7$		$\alpha_R = 0.8$	
$\alpha\beta$	−3.325	−3.64	3.8	4.16
$P(E < E_d)$	4.42×10^{-4}	1.36×10^{-4}		
$P(R \leq R_d)$			7.2×10^{-5}	1.6×10^{-5}

Table 4. Annual exceedance probabilities for effects and resistance at given target reliability index.

The Code provides formulas for deriving the design values of variables with different given probability distribution (normal, lognormal, Gumbel) that allows the use of the procedure to different type of hazards.

From the considerations above, it can also be concluded, that the annual probability of exceedance for the effects of actions 10^{4}/a will ensure the desired performance of SSCs. This conclusion can be interpreted in a way that the proper selection of design basis and design by conservative rules provide good chances to withstand the effects of "black-swan" earthquakes [54].

3.2. Evaluation of the design basis ground motion

During last three decades, a significant development can be observed in the methodology of the characterization of the seismic hazard for nuclear facilities. This has been motivated by the difficulties of assessing low probability earthquakes that have to be accounted for in the design basis. The return periods of these earthquakes exceeding the time span of historical records. Main focus of developments was to manage the uncertainty of characterization of low probability events that motivated the development of the probabilistic seismic hazard assessment (PSHA) methods. In addition to the inherent randomness of the phenomena, the PSHA methods take into account the uncertainty of modelling, i.e. epistemic uncertainty due to insufficient knowledge. In the development of PSHA methods, the document of the United States NRC NUREG/CR-6372 [55] published in 1977 was a milestone. The NUREG/CR-6372 defines four progressive levels of analyses in accordance with the completeness of information sources, complexity of analysis, but principally, with the systematic integration of diverse expert opinions. Its uppermost grade is Level 4 PSHA, which is based on the integration of expert opinions using sophisticated expert elicitation method. The practical preference is given to the Level 3 PSHA that also integrate the views of different experts, but with reasonable effort. This methodology is integrated by the document of NUREG/CR-6728 in 2001 [56], which covers the definition of the site response and provides guidance for defining the response spectrum of safe shutdown earthquake (SSE). Recently has been published a comprehensive study on lessons learned from application of the Level 3 and Level 4 SSHAC methodology [57]. Practical guidance to perform PSHA is given in the [58].

The scientific developments are also reflected in IAEA documents. The document 50-SG-S1 issued in 1979 was still fully based on the deterministic method. The IAEA guideline NS-G-3.3 issued in 2002 [59] already more particularly discusses the probabilistic method and pays greater attention to the quantification of uncertainties and to the elaboration of the seismological data base (microseismic monitoring, palaeoseismological examinations) as completely as possible. This tendency unambiguously manifests itself in the document NAÜ SSG-9 [34], which superseded the guideline NS-G-3.3 and recently in Revision 1 of the NS-R-3. There is also essential progress in the use of fault rupture modelling [60], and accounting the palaeo-seismicity in the seismic hazard evaluations [61].

The probabilistic method for defining seismic hazard consists of six fundamental steps:

1. Specification of seismic sources, source areas

2. Characterization of activity of seismic sources (magnitude-frequency-distribution, cut-off magnitude, depth distribution)

3. Selection of appropriate attenuation-lows that corresponds to the earth-physics features of the region

4. Development of logic tree and calculation of seismic hazard curve and definition UHRS at base-rock or outcrop

5. Accounting the site effects, calculation of the free field UHRS

6. Deaggregation of the hazard

Although the whole process is data driven, there is significant epistemic and aleatory uncertainty in all steps have:

• Knowledge of seismogenic potential of geological structures is not unambiguous and, consequently, therefore several seismotectonic models may exists;

• The knowledge on the activity of seismogenic structures is incomplete since the number of recorded events and the time span of observations is limited, consequently, beside of the instrumental and historical records, the palaeoseismic evidences have great importance [61];

• The attenuation-lows are based on the statistical processing of particular observations, but the sample number may be very small in areas where the number of strong earthquakes is low.

The logic tree is also suitable to consider the aleatory uncertainty, while the accounting of aleatory uncertainty can be managed by other methods, e.g. by the Monte-Carlo method.

The final output of the PSHA is the hazard curve, i.e. the annual probability of non-exceedance versus PGA and the UHRS at outcrop.

The seismic design basis and the hazard curve is generally defined as mean (black line) of possible realizations. The 14 and 84 percentiles (yellow and brown lines, respectively) and their deviations indicate the extent of the epistemic uncertainty.

Set of hazard curves are shown in **Figure 1** each of them corresponds to a branch in the logic tree.

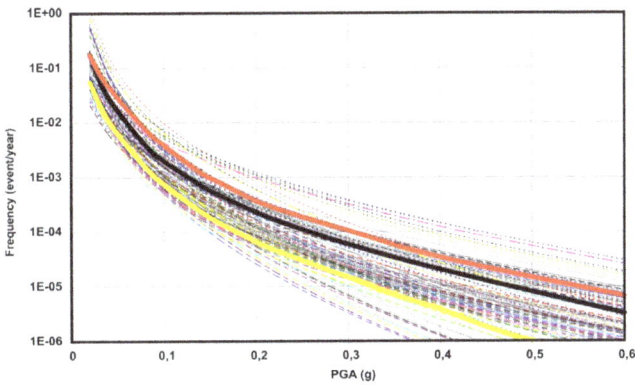

Figure 1. Hazard curves corresponding to branches of the logic tree.

Important result of the PSHA is the deaggregation of hazard that identifies the contributors to the hazard in magnitude-distance bins as it shown in **Figure 2**.

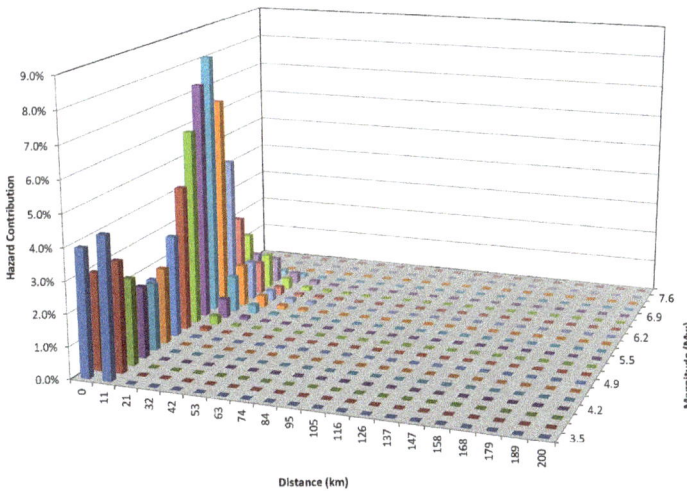

Figure 2. Deaggregation of 10^{-5}/a level PGA for a nuclear site.

Contrary to the PSHA, the deterministic method of seismic hazard evaluation accounts the aleatory uncertainties and the analysis is based on the only true model of seismogenic structures that is most supported by evidences. However, the selection of the only true model disqualifies all other views that might also be supported by observations.

A PSHA SSHAC Level 4 study has been made for Swiss NPP site that is well-published and widely discussed; see Refs. [62, 63].

3.3. Site effect, design basis response spectra

The SSE for the site is characterized by both horizontal and vertical free-field ground motion response spectra at the free ground surface. The procedure for calculating the free-field response is illustrated in **Figure 3**.

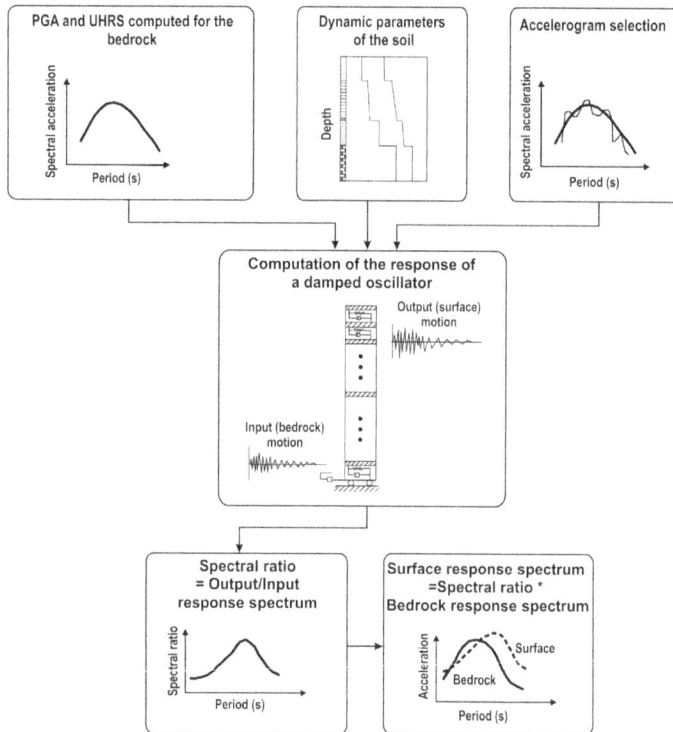

Figure 3. Calculation of the site-effect.

Ground motions at the foundation level and at the surface can then be computed, with account taken of the transfer functions of the overlying soil layers. The nonlinearity of the transfer to the surface can be essential for sites covered by soft soil layers. The ground response spectra can also be defined at the hypothetical outcrop.

In the designer practice, standardized response spectra are fitted to the free-field PGA; see for example the standardized response spectra given by Regulatory Guide 1.60 in Ref. [64]. The designers select an appropriate PGA for the basic/certified design.

In the United States, PGA = 0.3 g is selected for the new reactors to be certified. The peak ground acceleration of the SSE, referred as the Certified Seismic Design Response Spectra (CSDRS), has been established as 0.30 g for the AP1000 design [65]. These spectra are based on Regulatory Guide 1.60 with an increase in the 25 Hz region. The vertical peak ground acceleration is conservatively assumed to equal the horizontal value of 0.30 g. In Europe, the 0.25 g is set for the standardized response spectra [66]. The site-specific response spectra are used for adjusting the basic or certified design to the site and justification of margins.

3.4. Hazards caused by earthquakes

Among ground vibratory motion, earthquakes can cause surface settlement, permanent surface ruptures, landslides, soil liquefaction and tsunamis. The basic aspects of the surface rupture and soil liquefaction hazard will be presented below.

3.4.1. Evaluation of the design basis permanent surface movement

Surface deformation is distortion of geologic strata at or near the ground surface by the processes of folding or faulting as a result of various earth forces. Tectonic surface deformation is associated with earthquake processes. A fault shall be considered capable to cause permanent surface deformation, if on the basis of complex (geological, geophysical, seismological, palaeoseismological, geomorphological, etc.) investigations show the "evidence of past movement or movements (significant deformations and/or dislocations) of a recurring nature within such a period that it is reasonable to infer that further movements at or near the surface could occur" [34, 46].

The issue of surface rupture and capability/activity of faults at the sites has been addressed already in the "Introduction". In the earlier regulatory approach, the potential of surface movements was considered as absolute criterion for rejection of the site. The uncertainty, the definition of the surface rupture hazard was judged unacceptable and the possibility of engineering means for protection questionable.

The change of views can be tracked comparing the last three revisions of the United States NRC Standard Review Plan [67–69]. Recently, NRC regulations do not restrict building in an area with surface faulting potential, but if that potential exists, the regulations require that surface deformation must be taken into account in the design and operation of the proposed nuclear power plant [69]. Although it is not advised to locate a new plant at a capable fault, the issue seems to be unavoidable at the sites of some operating plants as it has been shown in the Introduction. Nowadays, finding a new nuclear site became rather difficult not because of scientific-technical difficulties but more because of political objections. The old nuclear sites are preferable to use for the location of new plants. Therefore, the analysis of surface fault hazard and consideration of the surface movement is a very burning issue.

The change of views on the surface movement issue reflects the development of the understanding both the phenomenon of surface rupture and the response of the plant to certain surface dislocation. In the IAEA Safety Guide SSG-9 [34], the probabilistic surface rupture hazard analysis is advised for the sites of operating plants, where is seems to be relevant.

According to the recent understanding of the issue, the judgement on the fault displacement hazard depends on the measure of displacement that would or would not challenge the safety. The new approach to the tectonic deformation hazard is based on the advances in the methodology to investigate and characterize the surface fault rupture and tectonic deformation achieved during last decades. In the same time, the engineering methods ensuring the required capacity of systems, structures and components to withstand certain amount of earthquake-induced ground deformations have been developed. These achievements are reflected in the new Japanese study [12] and United States standards [70]. The ANSI/ANS-2.30-2015 standard provides criteria and guidelines for assessing coseismic permanent ground deformation hazard due to tectonic surface fault rupture and deformation at nuclear facilities. The procedures and methods for performing probabilistic fault displacement hazard analysis for surface rupture hazard and probabilistic tectonic deformation hazard analysis for surface deformation due to displacements along blind (buried) faults are outlined in the standard. The logic of probabilistic surface displacement rupture hazard assessment is practically the same as the logic of the probabilistic seismic hazard analysis. The output of the fault displacement hazard is characterized by the hazard curve that shows probability of exceeding given level of displacement, i.e. $P(d > D) = 1.0 \, e^{(\lambda(z)T)}$, where the $\lambda(z)$ is the average frequency during time period T when the level of total displacement d exceeds D at the site resulting from earthquakes on all sources in the region. Z includes the contributions from principal- and distributed-fault displacements. The logic and the results of the probabilistic tectonic deformation hazard analysis are the same as in the case of performing probabilistic fault displacement hazard analysis. The ANSI/ANS-2.30-2015 standard references the relevant publications on the subject.

Rules for locating safety relevant structures relative to the near a fault are also advised in the standard. The procedure is as follows:

- Define permanent ground deformation zones on the basis of detailed geologic, geophysical, tectonic morphologic investigations.

- If the site is within the permanent ground deformation fault displacement analysis and/or tectonic deformation analysis has to be performed to understand the relationship of the building location relative to both the principal and distributed faults.

- Assess, whether the site is located within the permanent displacement zone greater than 200 m from principal fault zone and within the proximity (two times maximum foundation dimension) of distributed faulting, but no intersects the building foundation.

- Assess, whether the building is located within the permanent displacement zone greater than 200 m from principal fault zone, but the distributed faulting is within 200 m or directly intersects the foundation.

- Assess, whether the building is located inside the permanent displacement zone and within 200 m of principal faulting, but no distributed faulting directly intersects the foundation.

- Assess, whether the building is located inside the permanent displacement zone and within 200 m of principal faulting, and the distributed faulting directly intersects the foundation.

In all cases, the site can be accepted, if the results of the probabilistic fault displacement hazard analysis demonstrate that fault displacement hazard that is given in form of exceedance probability versus measure of displacement.

The basic concept and motivation of the Japanese study JANSI-FDE-03 rev.1 is to describe framework of the estimation of fault displacement to the assessment of plant safety against on-site fault displacement that account both the direct impact due to discontinuous displacement of the ground and indirect impact due to continuous deformation, such as inclination. The analysis methods have been developed primarily for application to existing plants, but the methods can also be referenced in the design of new plants. In the safety analysis, it is assumed that the secondary fault exists immediately beneath the reactor building and the possibility of its displacement cannot be denied, and its displacement will directly be applied to the reactor building. The design basis displacement δa can be performed via

- estimation by geological survey results on the basis of past displacements,

- estimation by analysis,

- probabilistic fault displacement hazard analysis, selecting the design basis displacement for the annual exceedance probability as in case of other external natural hazards.

The JANSI-FDE-03 rev.1 study references the relevant publications on the methods above.

Since the uncertainty of fault displacement is considered to be larger than that of other natural phenomena, the impact on the facilities arising from beyond design basis displacement δb has to be examined. In the study, it has been confirmed that secondary fault displacements for δb can be set to 30 cm with annual exceedance probability is less than 10^{-5} when the Mw = 6.5.

Deformation in the form of creep or after-slip and uplift and subsidence during subduction zone earthquakes is addressed for example in the Russian regulation seems to be consequent since it not allows to construct NPPs at the sites situated on active faults [71]. In this interpretation active are those faults along which relative displacement of the earth crust's adjacent blocks by 0.5 m and more took place during the last 1 million years (in the Quaternary period) [72]. (There are Russian normative documents where the attribute "5 mm/year recent movement" is also added, see, e.g. the RB-019-01 norm of the Rosatomnadzor.) The hazards categorized into three categories (hazard degrees) according to their severity measured by maximum allowable effects: high, moderate and low (denoted by I, II and III). The sites are also categorized into three Classes according to the category of site hazards (denoted by letters A, Б and В). According to this, the sites are of Class В, where a sudden movement along the fault (earthquake intensity 8 per MSK-64) is equal or larger than 0.3 m that is categorized as a Severity Degree I. For the creep and recent differential movements, the 0.3 m limit and additionally a larger then 10^6 m/a gradient of the Quaternary movement is set. In this case, considerations have to be made for selection another site. The site is acceptable if the displacement is less than 0.3 m and the velocity gradient of the Quaternary movement is less than 10^6 m/a. It is a remarkable coincidence that this limit is equal of those for beyond design basis displacement in the Japanese study JANSI-FDE-03 rev.1. Obviously, the limit value is related to the value of the limit tilt that can cause essential damages in reinforced concrete structures, that is approximately 0.003 in accordance with several standards. This limit is much less than the tilt of leaning tower of Pisa.

The standards ([34, 70, 72]) and the representative study [12] indicate that there are consolidated and applicable in the nuclear practice knowledge to deal with certain limited surface displacement. The very prudent position of the NRC reflects the reasonable position [69] stating that it is not forbidden building a nuclear power plant in an area with surface faulting potential, but if it is rather difficult to that the safety-related SSCs would maintain their safety functions if surface displacement occurs.

It has to be emphasized, the scientific-technical progress cannot be never braked by prudence, and it was and is always motivated by the needs and revelations. That is reflected in the sample of recent publications on the subject [73–80]. Special concerns regarding possibility to characterize the fault capability in moderate to low seismicity areas are discussed in Ref. [81].

3.4.2. Evaluation of liquefaction hazard

The secondary phenomena caused by earthquake have to be matter of acceptance of the site. According to the IAEA Safety Guide SSG-35 [44], if the potential for soil liquefaction is found to be unacceptable, the site shall be deemed unsuitable unless practicable engineering solutions are demonstrated to be available. Soil liquefaction is a ground failure or loss of strength that causes otherwise solid soil to behave temporarily as a viscous liquid. The phenomenon occurs in water-saturated unconsolidated soils affected by seismic waves. Poorly drained fine-grained soils such as sandy, silty and gravelly soils are the most susceptible to liquefaction. Since there are proven engineering solutions for soil stabilization and improvement against liquefaction hazard, the only site evaluation task is to locate the soil layers susceptible to liquefaction. The first screening can be performed on the basis of grain-size-distribution; see Ref. [82]. The screening can be performed calculating the factor of safety to the liquefaction, FS_{liq}, that is the measure of susceptibility that is the ratio of the CRR is the cyclic resistance-ratio, i.e. the available soil resistance to liquefaction expressed in terms of the cyclic stresses required to cause liquefaction, and CSR (cyclic stress ratio) that is the cyclic stress generated by the design earthquake.

The cyclic resistance-ratio CRR can be calculated as advised by US NRC Regulatory Guide 1.198 [83] on the basis of field and laboratory tests performed in accordance with Regulatory Guide 1.132 [84] and Regulatory Guide 1.138 [85], respectively. The liquefaction susceptibility can be calculated by analytical methods (effective stress method, linear and nonlinear), cyclic strength testing, physical modelling and empirical procedures. The latter are widely used in the practice. For CRR, empirical correlations have been developed on the basis of standard penetration tests (SPT), cone penetration tests (CPT), Becker penetration test (BPT) tests and on the shear wave velocity vs. field data. Since the statement on the liquefaction susceptibility is used for indication of the hazard and the design is focused on the complete avoidance, the analysis by empirical method is acceptable and sufficiently conservative.

In case of existing nuclear power plants at soil sites, the soil liquefaction can be a beyond design basis hazard that can cause design extension conditions. Safety analysis DEC case. Sophisticated methods have to be applied for the development of a best estimate assessment frame for evaluation of plant safety with respect to the liquefaction hazard; see Refs. [86–91]. Practical implementation for the Paks NPP site in Hungary can be found in Refs. [92–98]. A typical probabilistic liquefaction hazard analysis result, the annual probability of exceedance versus factor of safety, is shown in **Figure 4**.

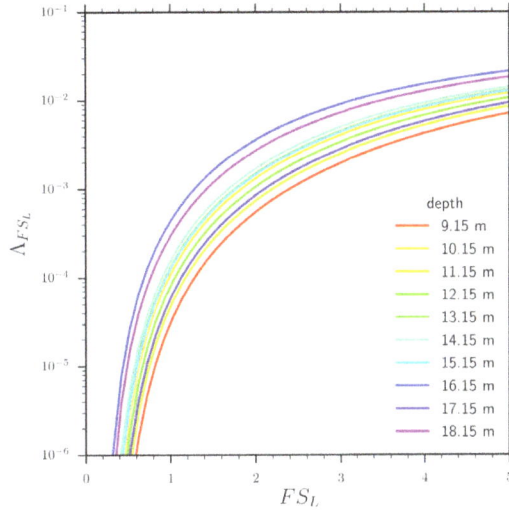

Figure 4. Annual probability of exceedance for factor of safety for different depth at a selected point of the site [94].

Practice shows that the different methods provide very scattering results because of differences in the modelling of the phenomena as it is shown in **Figure 5**; see Refs. [93, 94].

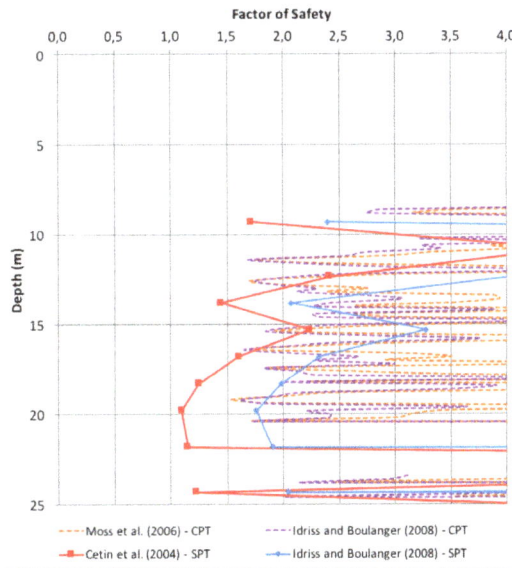

Figure 5. Factor of safety to liquefaction versus depth calculated by different methods [94].

The modelling uncertainties can be accounted for by logic tree. A method for dealing with epistemic uncertainty is proposed in Ref. [92] as it is shown in **Figure 6**.

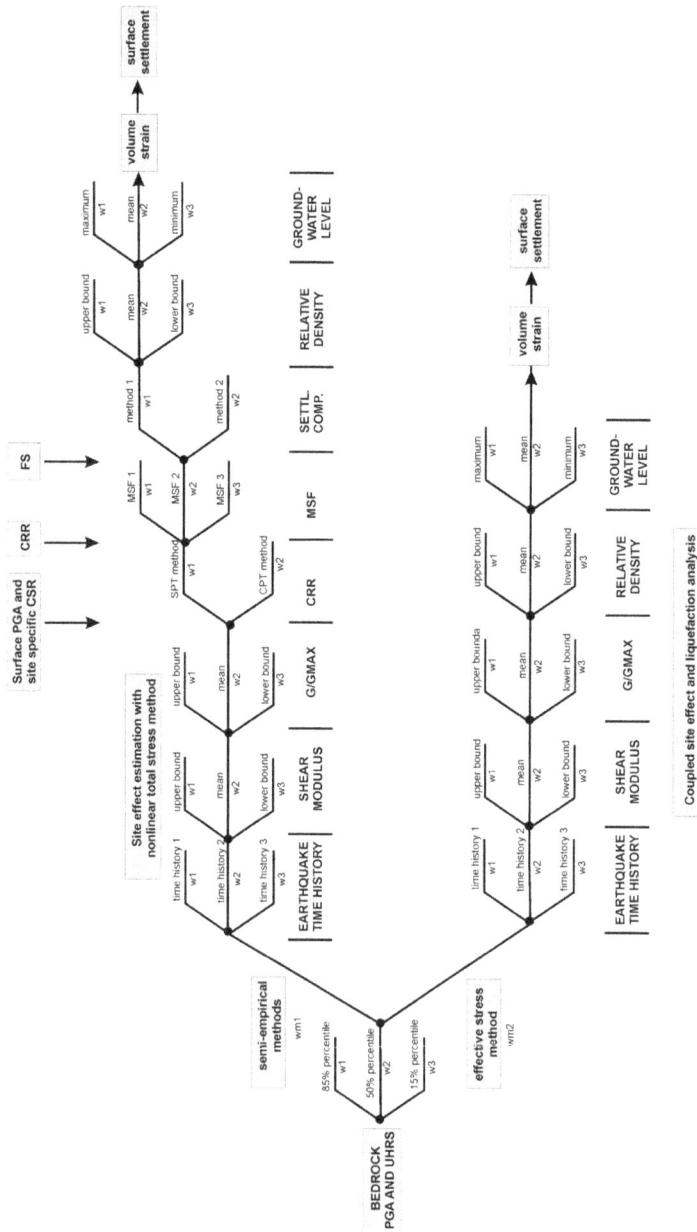

Figure 6. Logic tree elaborated for analysis of soil liquefaction [92].

3.5. Graded approach

When defining the design basis earthquake, it is practical to discriminate between nuclear facilities and reactors according to their potential risks [34]. The potential risk can be judged considering to the thermal output of the reactor, the quantity of activity stored, or on the basis of the characterization of the potential risk by means of 2nd or 3rd level PSA.

Designing the SSCs of a plant, the SSCs have to be classified according to the safety relevance. The design basis of the system or component can be defined in a graded way with the consideration of these two classifications. For example, in the case of classification into seismic-safety classes, the IAEA NS-G-1.6, the guideline for seismic-safe design, differentiates three seismic-safety and one "non-seismic-safety" classes or categories.

The ASCE/SEI 43-05 links the performance goal and hazard exceedance probability to the seismic design category that is assigned to the SSC, as a function of severity of consequence of the loss of function. There are five categories, SDC1 means conventional SSCs, while SDC5 is assigned to SSCs with high safety relevance. Target performance goal, PF and the HD is the average frequency of exceedance for the design basis earthquake is linked to the seismic design category and allowable of limit states A-D as it is shown in **Table 5**.

	SDC3	SDC4	SDC5
P_F	10^{-4}	4×10^{-5}	10^{-5}
H_D	4×10^{-4}	4×10^{-4}	10^{-4}

Table 5. Association of design category, target performance goal and exceedance frequency.

3.6. Regular review of the hazard evaluation

One of the most important lessons learned from the Fukushima-accident is the recognition of importance of regular review and updating of the site seismic hazard evaluation. According to the practice of many countries, the proper frame for the re-evaluation is the periodical safety reviews [99].

The regular assessment of experiences, feedback and periodical safety reviews, together with the safety enhancement measures, form an effective mechanism, which always guarantees the safety of the facility.

4. Basic design requirements

4.1. Generic design requirements and defence-in-depth

The design of the first and the second generation of nuclear power plants was governed by the first three levels of the defence in depth.

The classical two-earthquake-level design ensures continuous operation up-to OBE level exceeded. The design and qualification of practically all SSCs of the plant for OBE corresponds to the Level 1 of DiD. Installing automatic reactor, scram triggered by exceeding the OBE acceleration level corresponds to the Level 2 of DiD. The OBE could be considered as service-ability limit state for all non-safety-related SSCs of the plant.

Level 3.a had been ensured by design and qualification for SSE vibratory motion of SSCs needed for fundamental safety functions, the SSE is irreversible serviceability limit state for the SSCs ensuring fundamental safety functions.

The seismic safety re-evaluation and improvement programmes performed by many operators during last decades had been aimed to establish the compliance with the requirements of three levels of DiD.

The seismic PSA or seismic margin assessments performed used to quantify the seismic safety demonstrate that there are not cliff-edge effects since the margins designed in cover the unexpected beyond design basis earthquake effects. This was the Level 4 DiD in the earlier understanding.

The phenomena (tsunami, landslides, liquefaction) associated with the earthquake either had been considered as part of the design basis and appropriate protective measures had been implemented (for tsunami—breakwater walls, dykes, for liquefaction—soil improvement, appropriate foundation design) or had been excluded by proper site selection.

Extension of the design basis and accounting for the rare and severe external hazards are additional to the general design basis and represent more challenging or less frequent events. Consequently, in the new interpretation of defence-in-depth concept, design has to ensure protective measures for design extension conditions, i.e.:

- The SSCs that fulfil the basic safety functions have to be designed and qualified so that ensures extension of the capability due to predefined level of margin (Level 3b), or multiple protection (e.g. breakwater wall and tsunami guards). For these SSCs, the beyond design basis earthquake is irreversible limit state. The design extension conditions corresponding to Level 3.b differ from those corresponding to Level 3.a in the intensity of earthquake vibratory motion, but the response of the plant to the earthquake is as it is considered in the design basis.

- Means and procedures have to be in place for the case of very disastrous (black swan) earthquakes and associated with phenomena (Level 4). This can be correlated with ultimate limit state of all safety-related SSCs, except those designed or qualified specifically for DEC conditions (hardened core). Main technical objective is to maintain the integrity and leak-tightness of the containment. An example for a complementary safety feature is the equipment needed to prevent the damage of the containment due to combustion of hydrogen released during the core melt accident. The containment and its safety features shall be able to withstand extreme scenarios that include also the melting of the reactor core.

- Although the severe accidents have to be practically eliminated, means and plans have to be made for the case when the disastrous earthquakes result in severe core damage and releases (Level 5). Typical means and equipment are the bunkered and mobile equipment, and the emergency response centres rescue equipment rapidly available to support local operators.

The large scale common cause failures have to be assumed due to severe earthquakes that obviously enhance the possibility severe accidents. Us of redundant systems that is routine design solution for enhancing the reliability of safety functions does not provide additional safety improvement, since the earthquake affects simultaneously all redundancies that can result in common cause failures. More effective is the use of diverse systems and physical separation of safety systems. For example, in case of earthquake-induced fire, the physical separation can exclude the simultaneous loss of redundant safety systems. The systems dedicated for DEC conditions should be independent from the safety systems used as per design basis.

Phenomena, like soil liquefaction, generated by severe earthquake could be important for some sites as design extension situation.

The planning of severe accident management means and procedures and improvement of the disaster management on the country and international level became great attention [100].

The above design considerations have to be extended to the spent fuel pool, too.

4.2. Design for vibratory ground motion

Experience shows that the nuclear power plant design for the vibratory ground motion effects ensures is well established and conservative.

The IAEA SSR-2/1 [101] provides the general rules for design, particular rules for designing against earthquakes are given in Ref. [102]. Also, national regulations manage the question of safety protection against hazards in relation to level of importance; see, for example, the general design requirements of 10CFR50 Appendix A [46].

The national standards, for example, the ASME BPVC Section III [103], or the in the Russian standard NP-031-01 [104], specify the load combinations, the permissible stresses and the means of calculation of those stresses. The well-developed national standards are comparable [105].

A reasonable way to make the design effort rational is to apply the graded approach. The design basis and requirements with respect to the SSCs reliability can be defined in a graded way with respect to the safety classifications, see for example the IAEA safety guide NS-G-1.6 [102], that define three seismic-safety and one "non-seismic-safety" categories. For example, in the United States NRC regulation, it is reflected in 10CFR50 § 50.69 and Regulatory Guide 1.201 [106], which establish the basis of the categorization of SSCs according to safety relevance. The design codes provide graded approach in accordance with safety and seismic classes, see the Sections NB-3600, NC-3600 and ND-3600 according to the safety classes in the ASME BPVC III. Differentiation depending on the service level is also a conception element of design by "rules". Standards, for example, ASME BPVC Section III, unambiguously define the service level (design from Service Level A to D) into, which the given level is characterized by, which loads, and in a differentiated way defines the authoritative and permissible loads. The safety relevance is reflected in the selection of the design basis earthquake annual exceedance probability as well. Generally, the safety relevant SSCs have the seismic design basis 10^{-4}/a mean non-exceedance probability. The SSCs not classified have to be designed in accordance with industrial practice, i.e. for 475 return period earthquake as specified by the standard EUROCODE 8 [107].

As it is shown, the design for vibratory effects is well-established. The developments of design and analysis methodologies for vibratory ground motion are oriented on the reasonable decreasing of conservativism of the design methods. For example, accounting the incoherency phenomena in the soil-structure interaction [108–111] and development of performance-based design procedures.

4.3. Design for margins

The "design by rules" ensures margins that enable the SSCs to withstand the earthquake effects exceeding those in the design basis. The design codes ensure appropriate margins to compensate the inaccuracy of design methods, the uncertainty of loads and resistances the manufacturing and construction tolerances and defects, as well as the ageing effects.

For the unbiased quantification of the margin, the measure high confidence of low probability of failure (HCLPF) was introduced. High confidence of low probability of failure (HCLPF) is a measure of the seismic capacity of SSCs described in terms of a specified ground motion parameter (e.g. spectral acceleration) corresponding to 1% probability of unacceptable performance on a mean fragility curve. Otherwise, a given system component's failure probability associated to its HCLPF is lower than 5%, with a confidence of ≥95%.

The HCLPF is to calculate by conservative deterministic failure margin (CDFM) methodology, as given by the equation below, the whole procedure; see Ref [112]:

$$HCLPF = \frac{C - D_{NE}}{D_S + \Delta C_E} F_\mu a_{RLE} \qquad (6)$$

where a_{RLE} is the reference earthquake maximum horizontal acceleration, usually 0.3 g is selected (NUREG/CR-0098 response spectra fitted to the soil conditions at the site). The $C_E = C - D_{NE}$ is the part of C total capacity that is available to sustain the earthquake load since it is reduced by operating loads D_{NE}. The D_S is the earthquake load, and ΔC_{NE} represents the concurrent loads (for example, in a reinforced concrete reinforcing wall, tension simultaneously occurring with shearing). The capacity is defined according to the standards. The ductility is accounted by ductility factor F_μ or with the ductility-reduction factor $K_\mu = 1/F_\mu$. The calculation of system level and plant level HCLPF will be shown below, that needs modelling of systems and the success path of the plant for ensuring the fundamental safety functions.

According to the United States NRC requirements for new NPPs, the minimum acceptable plant level HCLPF is 1.67 times the design basis earthquake. For the new built, the European normative documents require a margin above the design basis equal to 1.4 times design basis peak ground acceleration. For existing plants, the HCLPF margin over the design basis earthquake is recommended as 1.4 according to the United States. Approach; see Ref. [113]. In other countries, this overall plant HCLPF is set to 1.5 [114].

Practical considerations regarding acceptable seismic margin of the severe accident management systems are published in Ref. [52].

4.4. Design for OBE

Operability of NPPs should be ensured after frequent but not severe earthquakes. The OBE is now a specific part of the design. Nowadays, the OBE is interpreted as an operational limit and inspection level rather than an obligatory design level. Definition of the OBE level is subject of owner considerations. Depending on the national regulations, an automatic reactor protection system has to be installed. The non-safety-related SSCs are designed for OBE that is usually selected in accordance with non-nuclear industrial standards. For example, in accordance with EUROCODE 8, the return period of OBE can be set to 475 years [107]. Selection of an OBE level higher than the design basis earthquake as per industrial standards will require additional design and qualification effort.

Design of safety-related SSCs for OBE level is not required, if the OBE PGA is equal or less than 1/3rd of the SSE PGA; see in Appendix S of the 10 CFR Part 50 in Ref. [115]. That exemption is to explain, if the allowable stresses in different service levels are compared to the allowable stresses for the Service Level D as these are defined by the ASME BPVC III.

4.5. Design of DEC provisions

Real challenge is the design DEC provisions since they have to survive the rare earthquakes and remain functional. Generally, it is allowed by the regulation that the SSCs that have to function under design extension conditions can be designed by realistic or best estimate methods. However, the term "realistic and best estimate" is not clearly specified in the regulations [101]. First step for developing a best estimate method is to identify the allowed design state. The ASCE/SEI 43-05 [49] categorizes the SSCs according to the maximum allowable deformation, i.e.:

- Large permanent distortion, short of collapse—significant damage

- Moderate permanent distortion—generally repairable damage

- Limited permanent distortion—minimal damage

- Essentially elastic behaviour—no damage

These damage categories can be assigned to the SSCs that are needed for different levels of DiD. The code defines the design procedure for each category. Proper definition of the design basis (Section 3.1.3) and design by rules (Sections 4.1–4.4, and 4.6) ensures the required function. The design issues are discussed and practical examples are given in Ref. [52]. There are specific facilities, systems and equipment needed for severe accident management, for example:

For protection of the containment

– Severe accident hydrogen management system

– In-vessel retention via external vessel cooling

– Containment venting

– Core catcher (to protect the containment from the molten core material)

Alternative power supply

– Autonomous power supply to designated consumers by mobile severe accident diesel generators

– Super-emergency diesel generator

Measurement and control systems

– Severe accident measurement system

Heat removal to the ultimate heat sink

– Alternative hear sink for reactor

– Spent fuel pool cooling system

Facilities that have to be available for severe accident management

– Protected command centre

– Backup command centre

– Barrack of fire brigade

These facilities, systems and equipment have to survive the rare earthquakes and be functional autonomously for period of time (e.g. 72 h) without support from outside.

4.6. Accounting for the ground surface displacement

As it has been discussed in Section 3.4.1, surface deformation can be accounted for in the design basis. Capable fault in the vicinity of the plant can cause sudden displacement below safety-related building structures that affect the strength of the foundation material and can result in deformation, inclination of the buildings and relative displacement between the buildings. In case of coseismic fault displacement, both the effects and consequences of vibratory motion and effects and consequence of fault rupture have to be taken into account. The slow tectonic movement can also cause tilting of the safety-related structures. Summary of numerous publications on the modelling of the behaviour of the structure under surface displacement/ deformation are given in Refs. [12, 70].

The study JANSI-FDE-03 rev.1 defines the following steps for analysis:

- Define the design basis displacement, δa.

- Perform analysis of coseismic displacement

- Perform analysis of deformation of foundation material

- Perform the design extension analysis for displacement δb exceeding the design basis one.

The analysis covers the building structures and the communication lines as well as the SSCs needed for fundamental safety functions. The analysis of the stability of foundation ground is also included into the scope.

The analysis of ground displacement and deformation as well as their consequences requires sophisticated ground-structure modelling and analysis techniques (finite element, discrete element methods, nonlinear, dynamic, etc.).

For the evaluation of integrity of the structures, the loads acting simultaneously have to be accounted for, i.e. the load arising from fault displacement δa, the dead-load, operating loads and seismic load that act together with the design basis displacement. For the design basis displacement δa, the allowable stresses and strains can be selected as for the ultimate conditions. In our understanding, the case is similar to the LS-B in accordance with ASCE/SEI 43-05 [49] or (or LS as per FEMA 365 [116]) conditions. For the beyond design basis displacement δb, the buildings and structures should not collapse LS-A in accordance with ASCE/SEI 43-05 or (or NC as per FEMA 365) conditions. These conditions can be evaluated also applying ASCE 41-03 or EUROCODE 8, Part 3 [117].

It has to be emphasized, that even in the case of DEC that corresponds to the Level 4 of DID, the desired condition of the containment would be to maintain the conditions as it is defined in the IAEA NS-G-1.10 Safety Guide [118]:

- Considering the structural integrity of the containment Level III: large permanent deformations. Significant permanent deformations and some local damages.

- For leak-tightness, the Level II condition, i.e. possible limited increase in leak rate. The leak rate may exceed the design value, but the leak-tightness can be adequately estimated and considered in the design.

5. Assessment of seismic safety

Safety analysis is a procedure that confirms compliance of the plant safety with the requirements and acceptance prescribed by nuclear regulations. The analysis is based on the physical modelling of the hazard effects and plant response via simulation and expert consideration. Response of power plant to earthquakes and judgement on the safety can be analysed by deterministic and probabilistic methods. These are as follows:

- Analysing whether the SSCs can withstand the earthquake effects using the design methods and justifying the code and the compliance for design basis earthquakes.

- Quantifying the margins, i.e. the load bearing and functional capacities of SSCs and the entire plant above design basis.

- Analysing whether the SSCs can withstand the effects of rare earthquakes (Level 4 of DiD) using specific rules.

- Evaluating core damage frequency due to earthquakes by probabilistic safety analysis, i.e. the seismic PSA is based on the modelling of the plant response to earthquake by event tree and fault trees. The seismic PSA includes the assessment of the containment function, too.

Obviously, the first method is nothing else as the design-by-rules. The third and fourth methods can already be considered as a routine one. Therefore, a brief information will only be given below. Most challenging seems to be the evaluation of the design extension conditions that will be discussed in some details.

5.1. Seismic margin analysis: SMA

The analysis and qualification of the margin against the impacts of hazards consists of the following main steps:

- Definition of a minimum configuration of SSCs that are needed for ensuring basic safety functions considering.

- Analysing the failure modes of the SSCs within the minimum configuration and calculation of HCLPF capacity of SSCs;

- The definition of the HCLPF capacity of the entire nuclear power plant.

The calculation of HCLPF by conservative deterministic failure margin (CDFM) methodology is given in Section 4.3. The rules for the calculation procedure and the applicable limitations are given in Ref. [112].

For calculation of HCLPF of a system, the fault tree of the system and its Boolean expression has to be developed. Let us consider a system consisting of two elements, A and B. Let us assume that for the function of the system survival of one of them is sufficient, i.e. an "or" relationship exists between the elements, and the advantageous output is C = A∪B if the elements are fully independent of each other, while in general $A \cup B = A + B - A \cap B$. If both elements are needed for the function of the system, i.e. an "and" relation exists between the elements, the success case is C = A ∩ B. Generalizing this procedure, it is possible to model of behaviour an arbitrarily complex system; see for example NUREG-0492 in Ref. [119]. The HCLPF of the systems needed for fundamental safety functions are calculated via Min-Max procedure. The Min-Max procedure seeks out both the weak link of the system (Min), if the elements are connected in series, and the strongest (Max) if the chains are connected in parallel with each other. Thus, the HCLPF of the system C consisting of elements A and B is as follows

$$HCLPF_C = Max\{HCLPF_A; HCLPF_B\} \tag{7}$$

if the model of system is as follows:

$$C = A \, or \, B = A \cup B = A + B, \tag{8}$$

respectively,

$$HCLPF_C = Min\{HCLPF_A; HCLPF_B\}. \tag{9}$$

For example, the Boolean representation of system E consisting of elements A, B, C and D is as follows

$$E = A*(B+C)*D. \tag{10}$$

The HCLPF capacity of the system is given as follows:

$$HCLPF_E = Max\{HCLPF_A; Min\{HCLPF_B; HCLPF_C\}; HCLPF_D\}. \tag{11}$$

The detailed description of the method, through the example of the qualification of the margin against earthquake, is included in NUREG/CR-4482 [120].

The HCLPF capacity defined for the nuclear power plant does not qualify core damage, but allows a conclusion on the likelihood that core damage will not occur (see for example ASME/ANS RA-S-2008 in Ref. [121]). In principle, it is possible, on the basis of the hazard curve, to assign the probability of exceedance or annual frequency, as well, to the impact corresponding to the HCLPF capacity.

5.2. Seismic PSA and PSA-based margin assessment

Seismic PSA contains three essential elements (see for example ASME/ANS RA-S-2008 in Ref. [121]):

1. assessment of the hazard in the form of a hazard curve showing the exceedance probability versus PGA,

2. a model of the power plant in the form of event and fault trees,

3. calculation of the failure rates using fatigue curves,

4. calculation of the conditional core damage frequency for the plant,

5. calculation of the core damage frequency of the plant.

The modelling of the plant top events and the seismically induced initiating events (IS) is illustrated in **Figure 7** taken from Ref. [122].

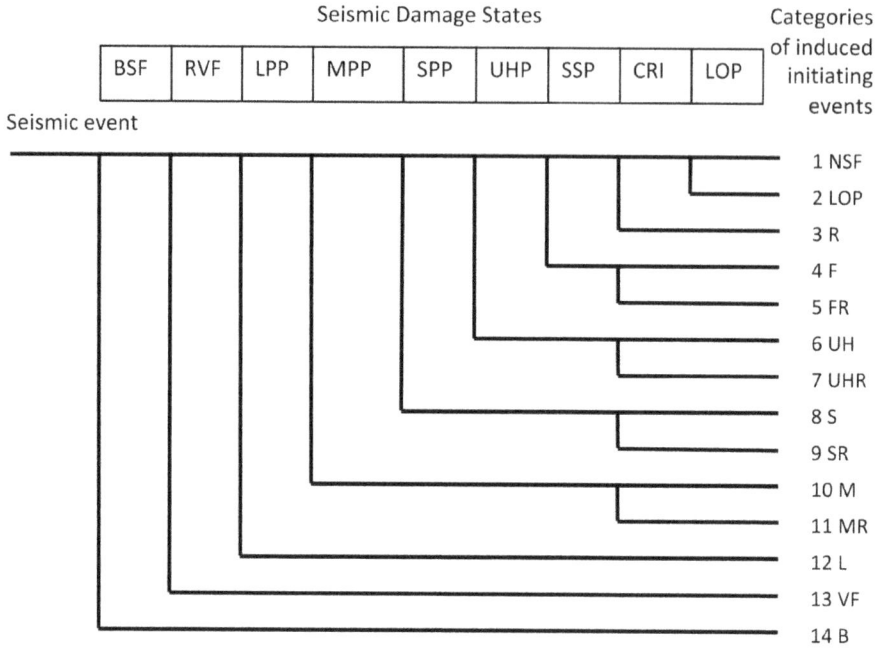

Figure 7. Modelling of seismically induced initiators and accident sequences with master event tree.

The concept of the probabilistic seismic safety assessment is rather simple. The probability of failure can be written in closed form, assuming that the hazard curve can be written as follows:

$$h(a) = k_0(a)^{-k} \tag{12}$$

where k0 is a constant and $k = 1/(lg(A_R))$, where A_R is the quotient of maximum horizontal accelerations in the case of a decrease in one decade of the probability of exceedance.

The conditional probability of structure or component to fail is assumed to be lognormally distributed:

$$P_f = P\left(fail\,|a \geq x\right) = \int_0^a \frac{1}{x\beta_C\sqrt{2\pi}} e^{\left(\frac{1}{2}\frac{\ln\left(x/C_m\right)^2}{\beta_C^2}\right)} \tag{13}$$

where $C_m = HCLPF^*e^{(2,326\beta_c)}$ is median capacity $\beta_C = (\beta_U^2+\beta_R^2)^{1/2}$ is standard deviation resulting from β_R randomness and β_U epistemic uncertainty. Here, however, β_U epistemic uncertainty is negligible, since it is negligible regarding susceptibility for damage. The total probability of failure is as follows:

$$P_{fail} = \int_0^\infty k_0\left(a'\right)^{-k} \frac{1}{a'\sqrt{2\pi}\,\beta} exp\left(-\frac{\left(\ln a' - \ln C_m\right)^2}{2\beta^2}\right) da' \tag{14}$$

Seismic margin assessment determined by a probabilistic method contains the second, third and the fourth elements of the above; therefore, the final result will be the conditional probability of core damage as a function of PGA [123]. Hence, the hazard assessment has rather high uncertainty, the capability of the plant ensured by the design is better to quantify by a method that is not integrate into the final result the uncertainty of site hazard evaluation.

Although the seismic PSA procedure is already standardized, further developments are needed for both modelling and fragility part of the PSA.

5.3. Practical example of the seismic margin analysis application

For a practical example showing the benefit of the seismic margin assessment is the case of the North Anna nuclear power plant (USA), on 23 August 2011, the plant was shaken by a magnitude 5.8, shallow-focus earthquake 11 miles away. Both units at the site have been shut down, and no damage compromising nuclear safety occurred. The PGA of the design basis earthquake of the power plant was 0.12 and 0.18 g (depending on the soil under the buildings); in contrast, the PGA of the actual quake was 0.26 g. A more than two-month-long supervision required more than 100.000 h of expert work and 21 million USD [124]. The units were restarted on November 11.

Figure 8 shows the response spectrum of the horizontal acceleration components of design basis earthquakes and operating basis earthquakes for containment base mat, as well as the response spectrum of the same acceleration components of the felt quake. As it can be seen, in the case of the response spectrum of the quake, the spectral amplitudes of the horizontal components exceeded the spectral amplitudes of the design response spectrum by 12%, while the amplitudes of the vertical acceleration response spectrum exceeded those of the design response spectrum by 21% on the average [125]. In addition, and that is what is essential for us, it also indicates the response spectrum of the deterministic SMA reference level earthquake for the base-mat. The exceedance of the design basis is unequivocal. It should also be noted that in the SMA calculation, the NUREG/CR-0098 response spectrum for 0.3 g PGA envelopes the response spectrum of the felt quake. In SMA calculations, only a few components produced HCLPF values lower than 0.3 g, but there was no failure in these cases, either. This is the first time when SMA could be practically and empirically tested and could be qualified as completely successful.

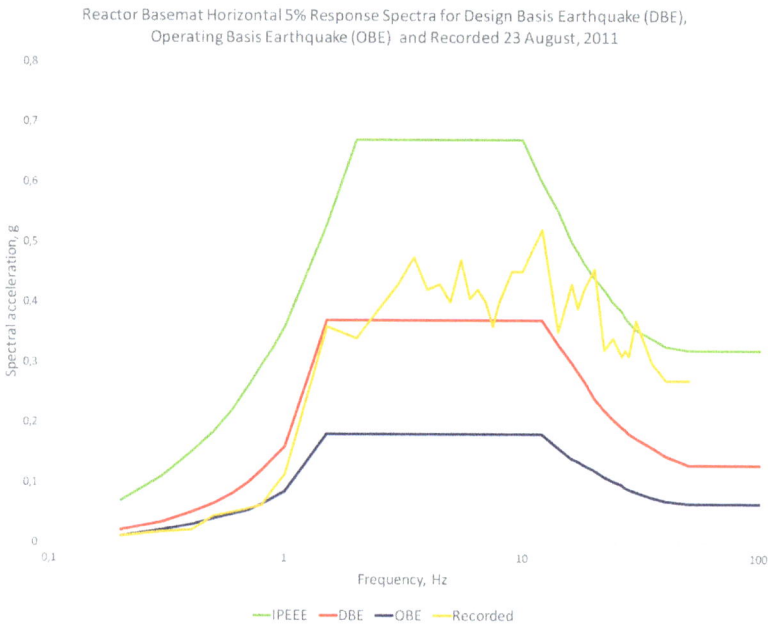

Figure 8. The response spectrum of the August 23 earthquake at North Anna nuclear power plant, compared to the response spectrum of the design basis (DBE), the operating basis (OBE), and the SMA referential (IPEEE) earthquakes [124].

A summary existing guidance on external hazard modelling is given in Ref. [126]. The limitations of deterministic and probabilistic safety assessments for external hazards are discussed in Ref. [127]. The findings are summarized in **Table 6**

The basic issues of the external hazard PSA including seismic PSA are identified in Refs. [128, 129]. The areas for further development are according to the latter study as follows:

1. External hazard screening/frequency assessment

2. Correlated hazards

3. External hazard impact assessment

4. Multi-unit sites

5. Mission time in Level-1 probabilistic safety assessments

6. Human reliability assessment for external hazards

7. Failure possibility for qualified equipment

8. Hydrogen explosion in the case of station black-out

9. Transient explosive materials in external event conditions

10. Connections between plant buildings and compartment

11. Spent fuel pool; waste treatment facilities

12. Modelling severe accident management guidelines

Examples of limitations in using current DSA —incomplete consideration of:	Examples of limitations in using current PSA —difficulties dealing with:
– Cliff-edges in terms of time for component operability while looking for success paths	– Consideration of limited mission time (typically 24 h)
– Feasibility of operator actions under conditions caused by extreme events	– Modelling of the impact of combined hazards on components and human actions
– Complex functional dependencies while looking for success paths	– Potential loss of important minimal cut sets (MCS) relevant for accidents caused by extreme events because those MCSs may be cut-off due to low probability
– Effects of combined hazards	– Large uncertainty probabilistic data may have
	– Usage of PSA tools by non-PSA safety analysts
	– Considerable time required for modelling
	– Difficulty to estimate the frequency of initiating events

Table 6. Limitations of the deterministic (DSA) and probabilistic (PSA) safety analysis methods while evaluating the plant safety in case of external hazards.

The actual needs for development of seismic PSA have been widely discussed in Refs. [130, 131]. Efforts made for extending the seismic PSA are reported, for example, in Ref. [123]. An attempt to develop alternative methodology is published in Ref. [127]. This research activity was aimed to develop a complementary analysis method to assess the robustness of the protection of nuclear power plants against extreme events and their combinations considering sufficiency of defence-in-depth provisions, including various dependencies, safety margins, application of specific design features, cliff-edge effects, multiple failures, prolonged loss of support systems and the capability of safety important systems for long-term operation. The method utilizes the qualitative information obtained from Level-1 internal initiating events probabilistic safety assessment studies (e.g. minimal cut sets), information on the operability limits of structures, systems and components and feasibility of operator actions under different severe conditions caused by extreme events. An advantage of the new method in comparison to traditional safety analysis is the direct consideration of combined load conditions resulting from the simultaneous occurrence of extreme external events.

5.4. Analysis for beyond design basis earthquake phenomena

A specific case of the safety analyses is the evaluation of plant post-event condition for the design extension conditions. Design extension condition would be that low probability event if a design basis earthquake causes soil liquefaction that was not considered in the design.

There are soil sites (for example at Paks NPP, Hungary) where this combination of events can have safety relevance [94, 96]. The deterministic analyses can be rather sophisticated, using coupled nonlinear soil-structure model for calculation of the settlement or differential settlement of the soil that is the input for evaluation of structural integrity. A probabilistic element is also present in the deterministic calculations hence the input parameters used for the calculation of settlement and soil-structure interaction are defined on a certain non-exceedance probability level and derived from the probabilistic seismic hazard assessment [93]. Mechanisms of soil deformations depend on the soil conditions, earthquake parameters and parameters of the structure in a very complex manner. This is shown in **Table 7**.

Increase in parameter	Primary deformation mechanisms/mechanism of displacement				
	Localized volumetric strains due to partial drainage	Sedimentation after liquefaction	Consolidation due to excess pore pressure dissipation	Partial bearing failure due to strength loss in the foundation soil	SSI-induced building ratcheting due to cyclic loading of foundation
PGA	↑↑	↑↑	↑↑	↑↑	↑↑
Rel. density	↓↓	↓↓	↓	↓↓	↑↓
Layer thickness	↑	↑	↑↑	↑	↑↓
Foundation width	↓	↑↓	↑	↓	↓↓
Static shear stress ratio	↓	↓	↓	↑↓	–
Ratio of height to width	↑	↑	↑	–	↑↑
Building weight	↑↓	↑↓	↑↓	↑↓	↑↑
3D drainage	↑↑	↓	↑	↓	↑↓

Table 7. Relation between mechanism of structural displacement and earthquake parameters as well as parameters of the structure.

It has been recognized, that the differential settlements and relative displacements between the different buildings and piping seems can be the major issue from the point of view of ensuring basic safety functions. This differential movement can be caused by slight variability of depth and thickness of the sediments. Therefore, settlements have to be regarded as the dominant engineering demand parameter.

A simplified event tree is shown in **Figure 9**. The loss of offsite power (LOSP) is assumed to be the initiating event. The reactor shutdown system (SCRAM system, denoted A) shall ensure the sub-criticality. The emergency power system (B) and the emergency core cooling system (C) are needed for avoiding the core damage. The success path after earthquake will be affected by the liquefaction with time delay Δt after strong motion starts. Some systems, once functioning during the earthquake, may not be affected by the liquefaction. For example, once dropped, the control rods will ensure the sub-criticality, though the reactor will be tilted due to the tilting of the reactor building that caused by the liquefaction.

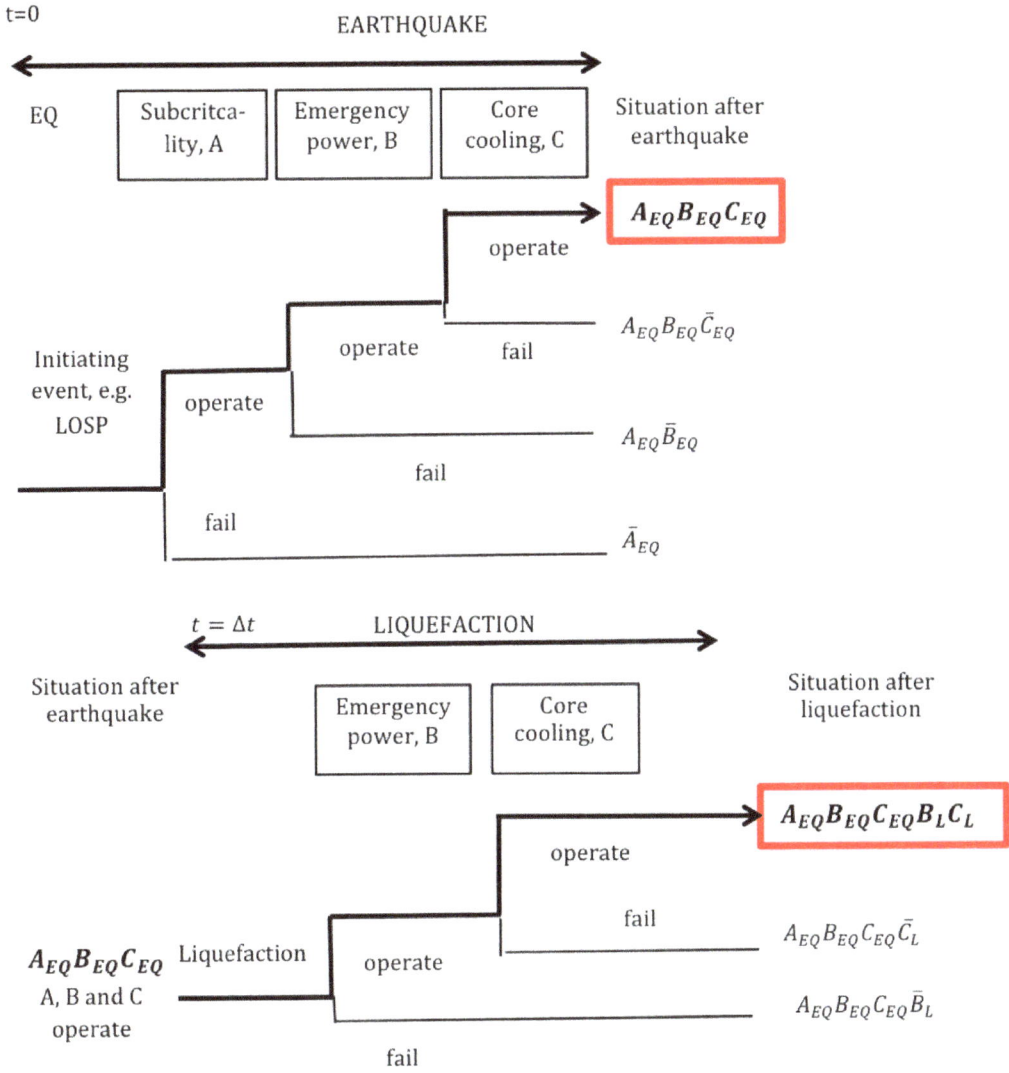

Figure 9. Simplified plant event tree for earthquake and liquefaction.

Although the time delay might be negligible and the liquefaction affects the ground motion at the site, it is reasonable to split the plant response and damages into two phases: response and damage to vibratory motion and response and damage to the liquefaction. The liquefaction is a separate load case subsequent to the vibratory motion, i.e. the plant structures do not feel the correlation between two phenomena.

Basis for the plant modelling is the model developed for probabilistic seismic safety assessment.

The specific deterministic safety analysis assesses the integrity and function of the plant structures consists of the following steps:

- Probabilistic seismic hazard assessment that provides the peak ground acceleration and deaggregation matrices that are used to perform probabilistic liquefaction hazard analysis. The PSHA provides the magnitude for the deterministic liquefaction hazard analysis.

- Calculation of soil settlements due to the liquefaction.

- Identification of SSCs within the scope of liquefaction safety analysis. These are the SSCs needed (or can be used) for heat removal from reactor and spent fuel pool and limitation of releases:

  Containment.

  SSCs that have to be functional or preserve their integrity:

  i.   The essential service water system that consists of piping, water intake structures and water intake control building. The underground pipelines connect the pumps located in the water intake building to the main reactor building and diesel building while crossing the lower level in the turbine hall. The intake channel is part of the system. Therefore, the slope stability has to be analysed.

  ii.   The emergency power supply system, the diesel buildings and their lifelines.

  iii.   The backup systems (e.g. the fire water system) that can be used as ultimate heat sinks in case of severe accident, and backup power supply systems, too.

  - Structures and systems with limited radioactive inventory (auxiliary building).

  - Part of the main building housing the control rooms.

  - Parts of the main building along the escape routes.

  - Barrack of fire brigade and protected command centre.

  - Laboratory and service building (workplaces and access to the controlled area)

  - Buildings that may collapse but should not damage the essential service water and emergency power systems or should not hinder the implementation of emergency measures.

- Definition of the desired condition of the structures from the point of view of safety and accident mitigation/management. Definition of the criteria for assessing whether the desired condition will be preserved. For example:

  a.   Permanent deformation of pipelines of the essential service water systems can be accepted assuming that the overall integrity and leak-tightness is ensured.

  b.   The water intake channel has not been blocked if the slopes slide down.

  c.   The safety of escape and access routes have to be ensured.

d. In case of containment the following conditions can be accepted; see Ref. [118]:

Structural integrity:

Level II: Local permanent deformations are possible. Structural integrity is ensured, although with margins smaller than those for design base.

Level III: Significant permanent deformations are possible, and some local damage is also expected. Normally, this level is not considered in case of severe accidents.

Leak-tightness:

Level II: The leak rate may exceed the design value, but the leak-tightness can be adequately estimated and considered in the design.

Considering the design of Paks NPP, large permanent deformations of the containment walls and floors are allowed when the deformations are within the strain limits allowable for the liner that ensures the necessary leak-tightness of the containment.

Relative displacement between containment building and other buildings has to be assessed from the point of view of integrity of essential service water pipelines crossing these locations.

e. The evacuation is ensured via safe escape routes. A near collapse condition is also acceptable in case of auxiliary building, but it has to preserve certain level of structural integrity for limiting the site radiation level.

• Development of the analysis methodology in line with graded approach, taking into account the importance of the structure and the features of the structure (e.g. foundation level compared to the depth of layers prone to liquefaction).

• Behaviour of the structures can be evaluated taking into account the foundation deformation due to dead weight of the building and additional foundation settlements due to liquefaction. In case of main reactor building, the soil settlement due to the liquefaction is affected by the static stress field. Otherwise, this effect can be neglected and the free-field settlement can be used as approximation of the deformation of foundations.

In all calculations, best estimate models and mean values of loads and material properties can be accepted. In best estimate models, some non-structural elements contribution to the resistance could be accounted. The calculation can be linear or nonlinear static. In case of containment (main reactor building), coupled soil-structure model is applicable.

• Evaluation of the integrity can be performed as it was shown in Section 4.6.

• Performing the analysis and definition of measures for accident mitigation/management.

Figure 10 shows the two extreme cases of the maps of soil settlements developed via CPT-based methodology of Zhang et al. in combination with method of Moss et al. and Robertson and Wride. The third option was the combination of methods Zhang et al. and Boulanger and Idriss. For independent control, the free surface settlement was computed using effective stress method to the average soil profile.

The analysis can demonstrate the availability of safety functions after soil liquefaction.

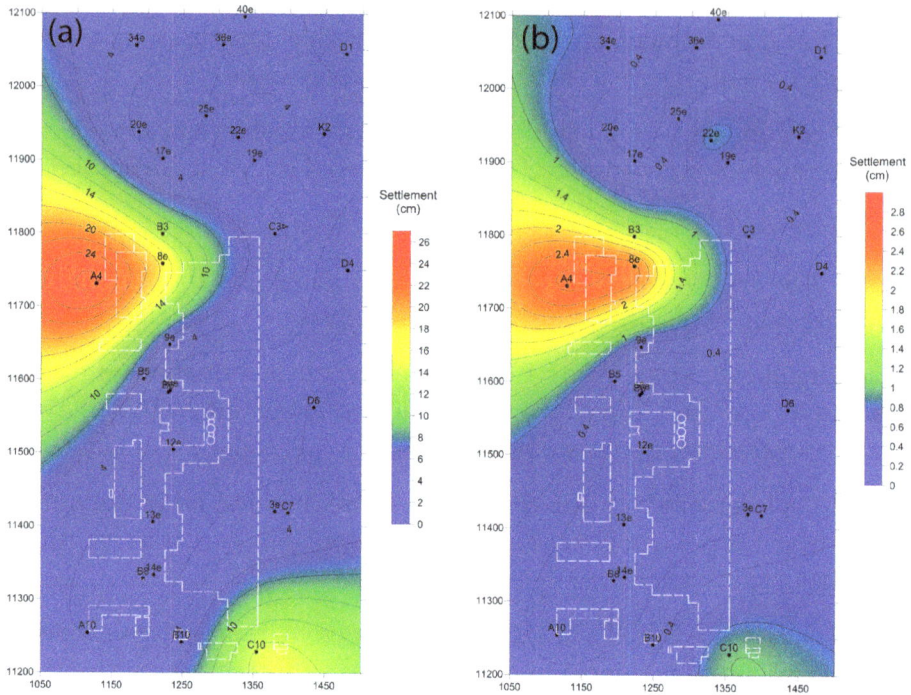

Figure 10. Maps of soil settlements at the site calculated by different methodologies.

6. Requirements for operation

6.1. Earthquake preparedness, procedures

The earthquake preparedness and post-earthquake procedures are well defined in the IAEA Safety Reports Series No 66 [132], Regulatory Guide 1.166 of the United States NRC [133],, documents of the EPRI [134, 135]. A very practical task for the operator is to maintain proper seismic housekeeping that is described in the EPRI document [136]. Here, some less emphasized but still important operational aspects of ensuring the earthquake safety are discussed.

6.2. Specific aspects of accident management

Proper design of the plant ensures that the SSCs that required for ensuring the plant safety remain functional both during and after the external event avoiding melting of the reactor core. The structures and systems required for accident management have to remain functional even in case of beyond design basis external events. The plant staff and the disaster management services of the country have to be prepared to manage extreme events and mitigate their consequences. This requirement has been formulated after the Fukushima accident and also adopted in the national regulations [37, 38].

The emergency planning and response requires evaluation of the consequences of external events beyond the scope of the plant design. A disastrous earthquake will cause catastrophic consequences in large area around the plant site. The post-event conditions around the site affect the logistical support of the emergency actions at the plant, influence the psychological condition of the plant personnel and determine the workload of the country's disaster management personnel.

In the paper [137], a hypothetical case study is presented analysing the consequences of a design basis earthquake for the region around a nuclear power plant of Paks Hungary. The aim of the study is to show, what would happen outside of the Paks Nuclear Power Plant, if a 10^{-4} annual probability earthquake would happen. In this case, the plant should be brought to safe shutdown condition. Although the plant safe shutdown is ensured, the plant personnel will need a minimum of logistical support from and communication with the outside area. Therefore, the results of the study can be used for planning of the logistical support of the plant accident management staff. The parameters (magnitude, focal depth and possible distance from the site) of the case-study earthquake are selected in accordance with the design basis of the Paks nuclear power plant). For evaluation of the damages of the built environment instrumental intensity map (shake-map) has been developed for the dominating the site seismic hazard earthquake. The distribution of population and housing data used in the study has been obtained by population census held in 2011 and published by the Hungarian Central Statistical Office. Based on these data, the damages have been assessed using European Macroseismic Intensity Scale and the corresponding phenomenological definition of damages. The intensity distribution in the affected by earthquake area is shown in **Figure 11**.

Serious damages of the unreinforced masonry and adobe dwellings in the area around the plant are expected. That affect the technical conditions for accident management and causes serious psychological load of the personnel doing the services.

There are several non-fixed loess slopes in the settlements and also along roads that are susceptible to sliding due to ground shakings. Sliding of the non-fixed loess slopes can block some roads. The damages of houses and lifelines could cause fires that are sometimes more severe than the effects due to vibratory ground motion.

In case of design basis earthquake, the electrical grid will suffer damages, since the towers of the grid have been designed for wind and ice loads assuming that the (100 years) earthquake loads are bounded with lateral wind loads. Due to the damages of the grid, nearly half of the domestic production will fall out from the power system that causes the collapse of the national grid. The Hungarian Independent Transmission Operator Company Ltd. has a recovery plan for the grid. Rebuilding of the grid could last from several hours to several days, depending on the severity of the damages. The NPP could stay in safe mode for minimum one week or unlimited long if the fuel supply for emergency diesel generators is continuous. It is also possible to operate one of the units on the reduced power level and ensure safe power supply for all units. The substation and the high-voltage towers at the NPP site have been upgraded for the design basis earthquake and these infrastructures will withstand the earthquake. Thus, the NPP will be a stable connection point for restoring the national grid as fast as possible. Since the NPP is the biggest producer in the country it is impossible to restore the grid without the NPP. If all units are out of operation, then for the restart for the first unit needs offsite power. Therefore, there are two independent transmission lines tested to transmit ~20 MW capacity to the NPP to restart the units. Consequently, the loss of power supply for the settlements around the plant will also worsen the situation for the people and make the work more difficult for the emergency services.

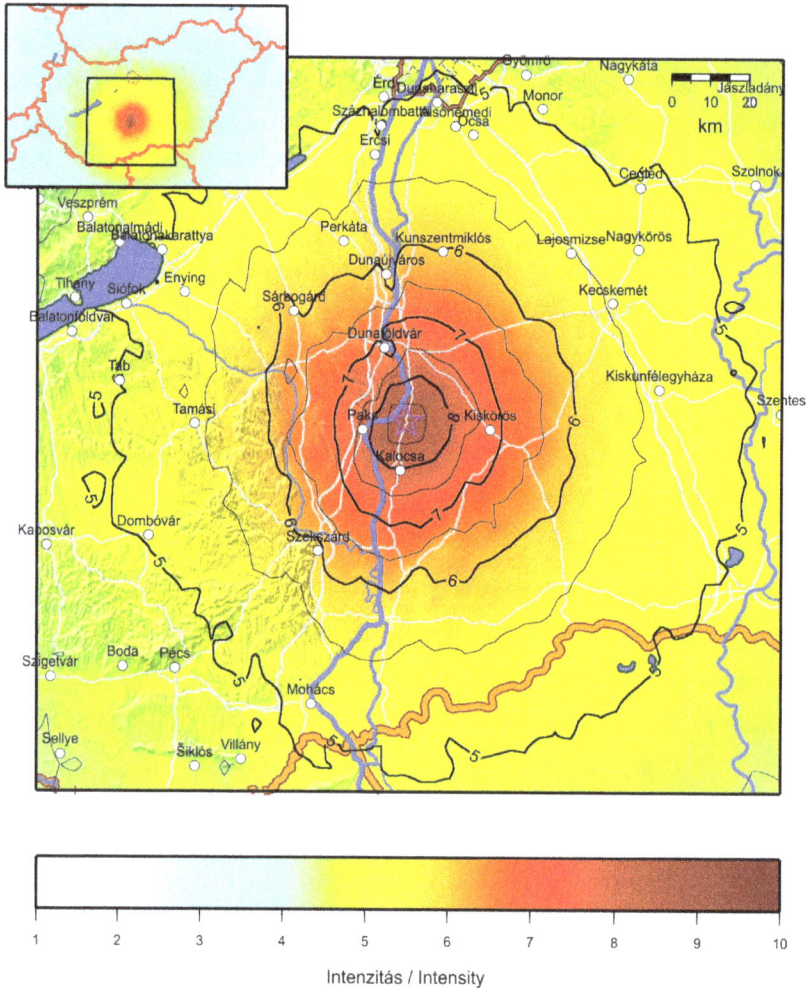

Figure 11. Instrumental intensity map of the selected scenario earthquake.

6.3. Restart after earthquake

After an earthquake, the condition of systems, structures and equipment at nuclear power plants has to be assessed since this information is needed for accident management and for the decision on the continuation of the operation.

There are two important issues to consider after an earthquake at a nuclear power plant: after a strong quake, the status of those systems that provide for basic safety functions has to be evaluated; while after a small quake, the conditions of restart, or (in case the plant remained in operation), the conditions of shutdown have to be determine. The first issue has been put forward by the tragic case of Fukushima, while the second has been the subject of intense investigations since the earthquake at the Kashiwazaki-Kariwa nuclear power plant.

The damaging potential of the earthquake can be characterized by maximum horizontal acceleration of the ground motion, response spectra, cumulative absolute velocity and different instrumental intensity values. These quantities can be correlated to the earthquake character-istics, magnitude, distance, etc. Some of the indices are selective to damage mechanism and can be correlated to load-bearing features of the structures. Different indicators of damaging potential of earthquakes are analysed in the paper [138, 139] from the point of view of applicability for post-event condition assessment at nuclear power plants and for using as criteria for restart of the operation.

A description of the procedure for post-event actions is given in Refs. [140, 141]. Traditionally, the basic damage indicator is the peak horizontal acceleration (PGA). As it has been shown in the Introduction, the plants can survive earthquakes with much larger PGA then those accounted for in the design. Consequently, the PGA cannot be considered as proper damage indicator.

The 23 August 2011 case of the North Anna nuclear power plant provided evidence not just for the adequacy of deterministically defined SMA, but also for the appropriateness of cumulative absolute velocity,

$$CAV = \int_0^T |a(t)| dt \tag{15}$$

as a failure indicator and failure avoidance criterion. Here, a(t) stands for the ground acceler-ation component, and T for the duration of the quake. When calculating standardized CAV, noise with an amplitude of ±0.025 g is filtered [142].

The criterion of failure avoidance or exceeding the OBE level is CAV ≥ 0.16 g s for any ground acceleration component. This criterion is not sufficient in itself, for it is also necessary to take into account the amplitude of the acceleration response spectrum calculated at 5% damping between 2 and 10 Hz, which has to be smaller than 0.2 g [140, 141]. In practice, the velocity criterion can be neglected (the spectral amplitude of velocity between 1 and 2 Hz should be ≥0.15 m/s). The CAV ≥ 0.16 g s criterion belongs to the failure conditions of structures not designed for earthquakes, with a large safety margin.

The adequacy of cumulative absolute velocity as a failure avoidance criterion has also been demonstrated by the case of the North Anna nuclear power plant [125]. The case has been also discussed in Ref. [139]. **Figure 12** shows the CAV values calculated for the components of measured acceleration, which can be compared with the CAV values of the design basis earthquake and the SMA reference-level quake. As can be seen, although the PGA of the 23 August 2011 quake exceeded the DBE PGA, on the basis of the CAV criterion there were no damages. This short, 25 s quake, the intensive phase of which only lasted for 3.1 s, obviously did not release significant energy, which is clearly indicated by the CAV. As the figure shows, the CAV of the quake is well below the CAV rendered to the DBE, while PGA and the response spectrum exceed it. In the light of this, it is no wonder that all this is majored by the CAV rendered to the reference level earthquake (RLE) used in SAM analysis.

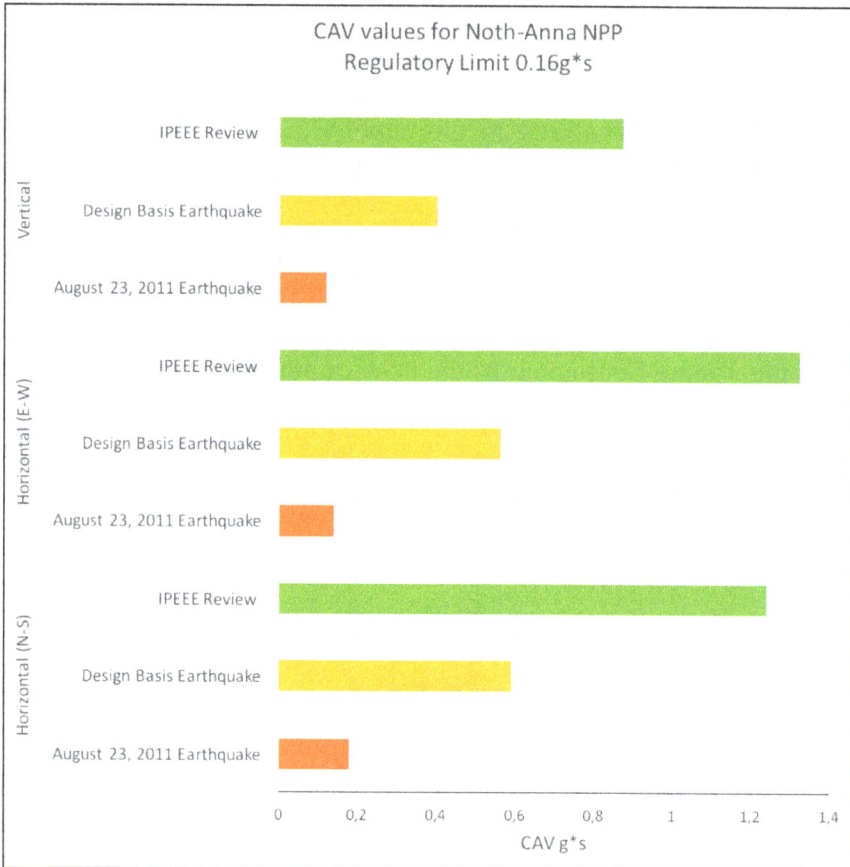

Figure 12. CAV of the 23 August 2011 earthquake at the North Anna nuclear power plant, and CAV's of the design basis event and the SMA RLE [125].

We may conclude from this that PGA and the response spectrum are highly conservative indicators of damage and damage avoidance. The DBE CAV is three times higher than the CAV of the actual quake, while the SMA RLE is almost nine or ten times higher than the CAV of the actual quake.

There are several studies published on the adequacy of the CAV as a parameter of energy input that can be well correlated to fatigue-type failure, see, e.g. the studies [143, 144]. The relation between the number of load cycles N with average amplitude Ac, during the T time of strong quakes and the CAV can be approximately written as follows:

$$CAV = \frac{1}{\pi}T \times A_c.$$ (16)

Due to the considerable amount of data available, the instrumental scale of the Japan Meteorological Agency IJMA is worthy of special interest. Another damage indicator can be MMI instrumental intensity, used by the US Geological Survey Shake Map. The Arias intensity is also an appropriate measure for damage. In Ref. [139], relations between these damage indicators are established and the similarity of the physical meaning is shown.

7. Conclusion

The brief presentation of the lessons learnt from Fukushima-accident and of discussion of selected issues of the seismic safety of nuclear power plants showed the enormous scientific and technical complexity of the subject. The ruggedness of the operating nuclear power plants against the earthquake vibratory motion is rather high. Nevertheless, the rare, devastating earthquakes are important contributors to the overall risk of nuclear power plant operation. It has also been demonstrated in the Chapter that the use of sophisticated and state-of-the art methods for hazard evaluation and design of the plant, safety against even the devastating earthquakes can be achieved. The experience of the Fukushima-accident changed the design philosophy. Instead of accounting of the effects of rather low probability earthquakes in the design basis, the design has to ensure the safety in case of rare earthquakes providing means and procedures for mitigation of core damage or if it is unavoidable for the mitigation of radioactive releases. Obviously, the responsibility of the designer is extended up-to the area of impossible. These new requirements have been discussed in the chapter, and the defence-in-depth concept is presented that is the basic approach to achieve the required level of safety. First of all, state-of-the-art methods have to be implemented for the site seismic hazard evaluation that includes both the characterization of the vibratory ground motion and the displacement due to surface rupture. The uncertainty of the hazard evaluation is the root cause of inadequate design basis definition. Therefore, the probabilistic hazard assessment methods are preferable in case of hazard evaluation that provides the possibility to quantify and manage the uncertainties. This approach has to be extended to the evaluation of hazards associated with earthquakes phenomena, e.g. soil liquefaction, surface displacement. As it is shown in the Chapter, the design has to be extended to the beyond design basis hazard levels, and the design basis extension conditions has to be accounted for in the design, safety analyses, as well as in the operator pre-earthquake preparedness. Analysis of design extension conditions that can be caused by beyond design basis liquefaction has shown the rather high ruggedness of the operating nuclear power plants.

The earthquakes affect not only the structures, systems and equipment of the plant, but the operators and all people responsible for the accident management. As it has been shown, the post-earthquake conditions at and around the site can be and essential factor influencing the staff while doing the service, and the external support of the accident management.

The safety of nuclear power plants also includes the safe continuous operation in case of a moderate earthquake and the safe restart after an earthquake. The selection and evaluation of appropriate damage indicators for assessing the post-earthquake plant condition is very important for both ensuring the safety and minimizing the economic losses. Finally, the benefits from nuclear power generation fairly compensate the risk, if the scientific-technical achievements are combined by high safety culture of the operators.

Author details

Tamás János Katona

Address all correspondence to: bata01@t-online.hu

1 University of Pécs, Pécs, Hungary

2 MVM Nuclear Power Plant Paks Ltd., Paks, Hungary

References

[1] OECD NEA. The Fukushima Daiichi Nuclear Power Plant Accident: OECD/NEA Nuclear Safety Response and Lessons Learnt, NEA No. 7161. OECD/NEA Publishing, Paris, 2013

[2] WENRA RHWG Report WENRA Safety Reference Levels for Existing Reactors—Update in Relation to Lessons Learned from TEPCO Fukushima Dai-Ichi Accident [Internet] 24th September 2014. Available from: http://www.wenra.org/media/filer_public/2016/07/19/wenra_safety_reference_level_for_existing_reactors_september_2014.pdf [Accessed: 2016-08-06]

[3] WENRA RHWG Report—Safety of new NPP designs, Study by Reactor Harmonization Working Group RHWG [Internet] March 2013. Available from: http://www.wenra.org/media/filer_public/2013/08/23/rhwg_safety_of_new_npp_designs.pdf [Accessed: 2016-08-06]

[4] WENRA Statement—Safety of new NPP designs [Internet] March 2013. Available from: http://www.wenra.org/media/filer_public/2013/04/05/wenra_statement_newdesigns2.pdf [Accessed: 2016-08-06]

[5] International Atomic Energy Agency, International Seismic Safety Centre [Internet], 2016. Available from: http://www-ns.iaea.org/tech-areas/seismic-safety/ [Accessed: 2016-08-06]

[6]   World Nuclear Association, Nuclear Power Plants and Earthquakes [Internet] July 2016. Available from: http://www.world-nuclear.org/information-library/safety-and-security/safety-of-plants/nuclear-power-plants-and-earthquakes.aspx [Accessed:2016-07-04]

[7]   IAEA Mission to Onagawa Nuclear Power Station to Examine the Performance of Systems, Structures and Components Following the Great East Japanese Earthquake and Tsunami, Onagawa and Tokyo, Japan, 30 July-11 [Internet] August 2012. Available from:http://www.iaea.org/inis/collection/NCLCollectionStore/_Public/44/050/44050829.pdf?r=1 [Accessed: 2016-07-04]

[8]   Y. Edano, B. Kaieda, G. Hosono (2011), Confirmation of the safety of nuclear power stations in Japan [Internet] July 11, 2011. Available from: http://japan.kantei.go.jp/incident/pdf/stresstest_e.pdf [Accessed: 2016-08-06]

[9]   IAEA (2007). Preliminary findings and lessons learned from the 16 July 2007 Earthquake at Kashiwazaki-Kariwa NPP, The Niigataken Chuetsu-Oki earthquake, Kashiwazaki-Kariwa NPP and Tokyo, Japan, 6–10 August 2007 [Internet], 2007. Available from:http://www.iaea.org/inis/collection/NCLCollectionStore/_Public/40/010/40010606.pdf [Accessed: 2016-07-04]

[10] WNN, Japanese stress test results approved, World Nuclear News [Internet] 13 February 2012. Available from: http://www.world-nuclear-news.org/RS-Japanese_stress_test_results_approved-1302124.html [Accessed: 2016-08-06]

[11] NRA, Nuclear Regulation Authority Enforcement of the New Regulatory Requirements for Commercial Nuclear Power Reactors, [Internet] July 8, 2013. Available from: http://www.nsr.go.jp/data/000067212.pdf [Accessed: 2016-08-06]

[12] JANSI-FDE-03 rev.1, Assessment Methods for Nuclear Power Plant against Fault Displacement, (Provisional Translation of Main Text), September 2013, On-site Fault Assessment Method Review Committee, Japan Nuclear Safety Institute. http://www.genanshin.jp/archive/sitefault/data/JANSI-FDE-03r1.pdf

[13] K.Okumura,(2010)Evaluationofnear-siteactivefaultsandeffectsonthesitebasedongeological structures. First Kashiwzaki International Symposium on Seismic Safety of Nuclear Installations, Kashiwazaki, Japan, 24–26 November 2010. http://www.nsr.go.jp/archive/jnes/seismic-symposium10/presentationdata/2_sessionA/A-14.pdf

[14] W. Epstein, S. Kotake, T. Yaegashi, A. Yamaguchi, (2013) Active faults and the restart of nuclear power plants in Japan: independent review team conclusions, 2013. PSA 2013 Session 314, Wednesday, September 25

[15] N. Chapman, K. Berryman, P. Villamor, W. Epstein, L. Cluff, H. Kawamura, (2014) Active faults and nuclear power plants. EOS, Transactions, American Geophysical Union. 95(4), 2014.

[16] Nuclear Engineering International, Be active [Internet] 4 March 2016. Available from: http://www.neimagazine.com/news/newsfaults-under-japans-shika-npp-said-to-be-active-4830171 [Accessed: 2016-07-29]

[17] World Nuclear News, Court rules against restart of Ohi reactors [Internet] 21 May 2014. Available from: http://www.world-nuclear-news.org/RS-Court-rules-against-restart-of-Ohi-reactors-2105146.html [Accessed: 2016-08-03]

[18] World Nuclear News, Tsuruga 2 sits on active fault, NRA concludes [Internet] 26 March 2015. Available from: http://www.neimagazine.com/news/newsfaults-under-japans-shika-npp-said-to-be-active-4830171 [Accessed: 2016-07-29]

[19] World Nuclear News, Safety review sought for Hamaoka 3 [Internet] 17 June 2015. Available from: http://www.world-nuclear-news.org/RS-Safety-review-sought-for-Hamaoka-3-1706154.html [Accessed: 2016-08-03]

[20] World Nuclear Association, Nuclear Power in Japan—Japanese Nuclear Energy [Internet] 21 July 2016. Available from:http://www.world-nuclear.org/information-library/country-profiles/countries-g-n/japan-nuclear-power.aspx [Accessed: 2016-07-29]

[21] R. D. Campbell, et al. (1998) Seismic re-evaluation of nuclear facilities worldwide: overview and status. Nucl. Eng. Des. 182:17–34

[22] NEA (1998) Status Report on Seismic Re-Evaluation. NEA/CSNI/R(98)5. OECD Publications, Paris

[23] A. Gürpinar, A. Godoy (1998) Seismic safety of nuclear power plants in Eastern Europe. Nucl. Eng. Des. 182(1): 47–58

[24] IAEA (2000) Benchmark study for the seismic analysis and testing of WWER Type NPPs, IAEA TECDOC 1176, IAEA, Vienna, October, 2000, ISSN 1011-4289

[25] T. J. Katona (2012) Seismic safety analysis and upgrading of operating nuclear power plants (Chapter 4). In: Wael Ahmed (ed) Nuclear power—practical aspects. InTech, New York, pp. 77–124. ISBN:978-953-51-0778-1

[26] G. Hardy, J. Kernaghan, J. Johnson, W. Schmidt (2008) EPRI independent peer review of the TEPCO seismic walkdown and evaluation of the Kashiwazaki-Kariwa nuclear power plants: a study in response to the July 16, 2007, NCO Earthquake. EPRI, Palo Alto. 1016317.

[27] N. C. Chokshi, Y. Li, V. Graizer. Lessons learned from post-earthquake investigations at North Anna Nuclear Power Plant. Presented at International Workshop on "Safety of Multi-Unit NPP Sites against External Natural Hazards", Mumbai, India, October 17–19, 2012, [Internet] 2012. Available from: https://issc.iaea.org/show_Document.php?id=1196, [Accessed: 2012-10-24]

[28] NRC, A comparison of U.S. and Japanese regulatory requirements in effect at the time of the Fukushima accident, Report. [Internet] November 2013. Available from: http://www.nrc.gov/docs/ML1332/ML13326A991.pdf [Accessed: 2016-08-04]

[29] NRC, Recommendations for Enhancing Reactor Safety in the 21st Century, The Near-Term Task Force Review of Insights from the Fukushima Dai-Ichi Accident [Internet] July 12, 2012. Available from: http://www.nrc.gov/docs/ML1118/ML111861807.pdf [Accessed: 2016-08-04]

[30] EPRI, Seismic Evaluation Guidance: Screening, Prioritization and Implementation Details(SPID) for the Resolution of Fukushima Near-Term Task Force Recommendation 2.1: Seismic, 1025287 [Internet] 28-Feb-2013. Available from: http://www. epri.com/abstracts/Pages/ProductAbstract.aspx?ProductId=000000000001025287 [Accessed: 2016-08-06]

[31] NRC Regulatory Guide 1.208, A Performance-Based Approach to Define the Site-Specific Earthquake Ground Motion, U.S. Nuclear Regulatory Commission, March 2007

[32] EPRI, Seismic Evaluation Guidance, Augmented Approach for the Resolution of Fukushima Near-Term Task Force Recommendation 2.1—Seismic [Internet] 31 May, 2013. Available from: http://www.epri.com/abstracts/Pages/ProductAbstract.aspx?ProductId=000000003002000704 [Accessed: 2016-08-06]

[33] Commission Staff Working Paper, Technical Summary of the national progress reports on the implementation of comprehensive risk and safety assessments of the EU nuclear power plants, European Commission, Brussels, 24.11.2011 SEC (2011) 1395 final

[34] Seismic Hazards in Site Evaluation for Nuclear Installations: Safety Guide No. SSG-9, International Atomic Energy Agency, Vienna, 2010. ISBN 978–92–0–102910–2

[35] Safety Standards Series NS-G-1.6, Seismic Design and Qualification for Nuclear Power Plants, IAEA, Vienna, 2003

[36] Slovenian National Report on Nuclear Stress Tests, Final Report, December 2011, Slovenian Nuclear Safety Administration. http://www.ensreg.eu/sites/default/files/Slovenian%20Stress%20Test%20Final%20Report.pdf

[37] WENRA RHWG, Guidance Document Issue T: Natural Hazards, Head Document, 18 February 2015

[38] WENRA RHWG, Guidance on Safety Reference Levels of Issue F: Design Extension of Existing Reactors, 29 September 2014

[39] Nuclear Power in Japan, World Nuclear Association, 21 July 2016. http://www.world-nuclear.org/information-library/country-profiles/countries-g-n/japan-nuclear-power.aspx

[40] Nuclear Power in the World Today, World Nuclear Association. http://world-nuclear.org/information-library/current-and-future-generation/nuclear-power-in-the-world-today.aspx

[41] IAEA Fundamental Safety Principles: Safety Fundamentals, IAEA Safety Standards Series, No. SF-1, International Atomic Energy Agency, Vienna, 2006. ISBN 92–0–110706–4

[42] Government Degree on the Safety of Nuclear Power Plants 717/2013 Translation from Finnish. Ministry of Employment and the Economy, Finland

[43] U.S. NRC 10CFR Part 50, Domestic Licensing of Production and Utilization Facilities (NRC, 1956).

[44] IAEA, Site Survey and Site Selection for Nuclear Installations, IAEA Safety Standards Series No. SSG-35. International Atomic Energy Agency, Vienna, 2015. ISBN 978–92–0–102415–2

[45] IAEA Site Evaluation for Nuclear Installations, IAEA Safety Standards Series No. NS-R-3 (Rev. 1), International Atomic Energy Agency, Vienna, 2016. ISBN 978–92–0–106515–5

[46] US Nuclear Regulatory Commission (NRC) 10 CFR 100, Reactor Site Criteria, Appendix A. Seismic and Geologic Siting Criteria for Nuclear Power Plants. http://www.nrc.gov/reading-rm/doc-collections/cfr/part100/part100-appa.html

[47] Accounting of External Natural and Man-Induced Impacts on Nuclear Facilities, NP-064-05, Federal Environmental, Industrial and Nuclear Supervision Service, Moscow, 2005. http://en.gosnadzor.ru/framework/nuclear/NP-064-2005.pdf

[48] Nuclear Power Plant Siting–Main Criteria and Safety Requirements, NP-032-01, Gosatomnadzor, 8 November 2001, [Internet] 2002. Available from: http://www.gosthelp.ru/text/NP03201Razmeshhenieatomny.html [Accessed: 2016-08-08]

[49] ASCE/SEI 43-05 Seismic Design Criteria for Structures, Systems, and Components in Nuclear Facilities, Published by American Society of Civil Engineers, 1801 Alexander Bell Drive, Reston, Virginia 20191. ISBN 0-7844-0762-2

[50] Guide YVL B.7, Provisions for Internal and External Hazards at a Nuclear Facility, STUK, [Internet] 2013. Available from: http://www.finlex.fi/data/normit/41791-YVL_B.7e.pdf [Accessed: 2016-08-08]

[51] Regulatory Guide 1.208 A Performance-Based Approach to Define the Site- Specific Earthquake Ground Motion, U.S. NRC, March 2007

[52] T. J. Katona, A. Vilimi (2016) Design of Severe accident management systems for beyond design basis external hazards at Paks NPP. Proceedings of the 2016 24th International Conference on Nuclear Engineering, ICONE24, June 26–30, 2016, Charlotte, North Carolina, paper ICONE24-60939

[53] Eurocode EN 1990:2002: Basis of structural design, [Internet] 2002. Available from: https://www.unirc.it/documentazione/materiale_didattico/599_2010_260_7481.pdf [Accessed: 2016-08-22]

[54] J-U. Klügel, Consideration of "Black Swan" Events in the Seismic Safety Review and the Seismic Upgrade Program of Existing Nuclear Power Plants—the NPP Gösgen Example, Post-SMiRT23 Seminar, Istanbul, Turkey, October 21–23, 2015

[55] NUREG/CR-6372, SSHAC—Senior Seismic Hazard Analysis Committee Report: Recommendations for Probabilistic Seismic Hazard Analysis: Guidance on Uncertainty and Use of Experts, U.S. NRC, 1997

[56] NUREG/CR-6728 Technical Basis for Revision of Regulatory Guidance on Design Ground Motions: Hazard- and Risk-consistent Ground Motion Spectra Guidelines, Prepared by R. K. McGuire, W. J. Silva, C. J. Costantino, U.S. NRC 2001.

[57] NUREG-2117, Rev. 1, Practical Implementation Guidelines for SSHAC Level 3 and 4 Hazard Studies. Prepared by A. M. Kammerer and J.P. Ake, U.S. NRC, 2012

[58] ANSI/ANS-2.29-2008 Probabilistic Seismic Hazard Analysis. American National Standards Institute/American Nuclear Society. 2008, Le Grange Park, Illinois.

[59] Evaluation of Seismic Hazards for Nuclear Power Plants: Safety Guide No. NS-G-3.3, International Atomic Energy Agency, Vienna 2002. ISBN 92–0–117302–4

[60] Ground motion simulation based on fault rupture modelling for seismic hazard assessment in site evaluation for nuclear installations, International Atomic Energy Agency, Vienna Safety Reports Series No. 85. ISBN 978–92–0–102315–5

[61] The contribution of palaeoseismology to seismic hazard assessment in site evaluation for nuclear installations, IAEA-TECDOC series No. 1767, International Atomic Energy Agency Vienna, 2015. ISBN 978–92–0–105415–9

[62] Ph. L. A. Renault, N. A. Abrahamson, K. J. Coppersmith, M. Koller, Ph. Roth, A. Hölker (2014) PEGASOS refinement project, probabilistic seismic hazard analysis for Swiss Nuclear Power Plant Sites, vol 1. Summary Report©2013-2015 swissnuclear, Olten, 20. December 2013 Rev.1: 20. December 2014

[63] P. Renault (2014) Approach and challenges for the seismic hazard assessment of nuclear power plants: the Swiss experience, Bollettino di Geofisica Teorica ed Applicata. 55(1): 149–164. doi:10.4430/bgta0089

[64] Design Response Spectra for Seismic Design of Nuclear Power Plants, Regulatory Guide 1.60, Revision 2, U.S. Nuclear Regulatory Commission, July 2014

[65] AP1000 Design Control Document, 3.7 Seismic Design. http://www.nrc.gov/docs/ML1117/ML11171A430.pdf

[66] EUR, European Utility Requirements for LWR Nuclear Power Plants, vol 2, Chapter 1, Safety Requirements (Part 2)

[67] NUREG-0800 Standard Review Plan, 2.5.3 Surface Faulting U.S. Nuclear Regulatory Commission, Revision 3, March 1997

[68] NUREG-0800 Standard Review Plan, 2.5.3 Surface Faulting U.S. Nuclear Regulatory Commission, Revision 4, March 2007

[69] NUREG-0800 Standard Review Plan, 2.5.3 Surface Faulting U.S. Nuclear Regulatory Commission, Revision 5, July 2014

[70] ANSI/ANS-2.30-2015: Criteria for Assessing Tectonic Surface Fault Rupture and Deformation at Nuclear Facilities, American Nuclear Society, La Grange Park

[71] Nuclear Power Plant Siting. Main Criteria and Safety Requirements, NP-032-01, Russian Gosatomnadzor, Moscow, 2002. http://en.gosnadzor.ru/framework/nuclear/NP-032-2001.pdf

[72] Accounting of External Natural and Man-Induced Impacts on Nuclear Facilities, NP-064-05, Rostekhnadzor of Russia, Moscow 2005. http://en.gosnadzor.ru/framework/nuclear/NP-064-2005.pdf

[73] Paul C. Rizzo Associates, Inc. (2013) Probabilistic fault displacement hazard analysis Krško East and West Sites, Proposed Krško 2 Nuclear Power Plant, Krško, Slovenia, Revision 1, Technical Report, Project No. 11-4546, 13 May 2013. http://www.ursjv.gov.si/fileadmin/ujv.gov.si/pageuploads/si/medijsko-sredisce/dopisGen/PFDHA_Studija.pdf

[74] Paul C. Rizzo Associates, Inc. (2013) Sensitivity Analysis, Probabilistic Fault Displacement Hazard Analysis Krško East and West Sites, Proposed Krško 2 Nuclear Power Plant", Krško, Slovenia, Revision 1, Final Technical Report, Project No. 11-4546, 31 May 2013. http://www.ursjv.gov.si/fileadmin/ujv.gov.si/pageuploads/si/medijsko-sredisce/dopisGen/R4_Final_Technical_Report-Sensitivity_Analysis_PFDHA_Rev._1.pdf

[75] Y. Suzuki (2014) Fault Displacement Assessment Methods. Japan Nuclear Safety Institute (JANSI) for Nuclear Power Plant, Plenary Meeting of the International Seismic Safety Centre's Programme, Vienna, Austria, 27–31 January 2014. Conference ID:47353 (TM)

[76] J. Johansson (2014) Fault induced ground deformation: comparison of field and experimental observations with numerical simulations. Plenary Meeting of the International Seismic Safety Centre's Programme, Vienna, Austria, 27–31 January 2014. Conference ID:47353 (TM)

[77] A. Gürpinar (2014) New approaches in addressing fault displacement hazard at NPP Sites. Fault displacement hazard meeting, ISSC, International Atomic Energy Agency, January 27–31, 2014

[78] Mark D. Petersen (2014). United States Engineering Treatment for Assessing Tectonic Surface Fault Rupture and Deformation, USGS National Seismic Hazard Mapping

Project. Fault displacement hazard meeting, ISSC, International Atomic Energy Agency, January 27–31, 2014

[79] Paul C. Rizzo (2015) Evaluation of External Hazards-Consideration of Geologic Faults. Post-SMiRT23 Seminar, Istanbul, Turkey October 21–23, 2015

[80] A. Gürpınar, L. Serva (2015) A risk informed engineering approach to consider faults near an NPP. Transactions, SMiRT-23, Manchester, United Kingdom, August 10–14, 2015, Division IV, Paper ID 423

[81] I. Prachař (2015) Identification of suspicious faults in areas of low seismicity—an example of the Bohemian Massif, best practices in physics-based fault rupture models for seismic hazard assessment of nuclear installations, Vienna, Austria, November 18–20, 2015. https://www.researchgate.net/publication/296195485

[82] KTA 2201.2 (2012–11), Design of Nuclear Power Plants Against Seismic Events Part 2: Subsoil. Safety Standards of the Nuclear Safety Standards Commission (KTA). http://www.kta-gs.de/d/regeln/2200/2201_2_r_2012_11.pdf

[83] Regulatory Guide 1.198. Procedures and Criteria for Assessing Seismic Soil Liquefaction at Nuclear Power Plant Sites, U.S. NRC, November 2003

[84] Regulatory Guide 1.132. Site Investigations for Foundations of Nuclear Power Plants. Revision 2, U.S. NRC, October 2003

[85] Regulatory Guide 1.138. Laboratory Investigations of Soils for Engineering Analysis and Design of Nuclear Power Plants. U.S. NRC, April 1978

[86] NUREG/CR-6622. Probabilistic Liquefaction Analysis, USNRC, November 1999.

[87] Youd és Idriss (2001) Liquefaction resistance of soils: summary report from the 1996 NCEER and 1998 NCEER/NSF workshops on evaluation of liquefaction resistance of soils

[88] R. B. Seed, K. O. Cetin, R. E. S. Moss, A. M. Kammerer, J. Wu, J. M. Pestana, M. F. Riemer, R. B. Sancio, J. D. Bray, R. E. Kayen, A. Faris (2003) Recent advances in soil liquafaction engineering: a unified and consistent framework. Report No. EERC 2003-06. University of California, Berkeley

[89] R. T. Mayfield (2007) The return period of soil liquefaction. PhD thesis, University of Washington, p. 292.

[90] S. L. Kramer, R. T. Mayfield (2007) Return period of soil liquefaction. J. Geotech. Geoenviron. Eng. 133(7):802–813

[91] K. Goda, G. M. Atkinson, J. A. Hunter, H. Crow, D. Motazedian (2011) Probabilistic liquefaction hazard analysis for four Canadian cities. Bull. Seismol. Soc. Am. 101:190–201. doi:10.1785/0120100094

[92] E. Győri, L. Tóth, Z. Gráczer, T. Katona (2011) Liquefaction and post-liquefaction settlement assessment—a probabilistic approach. Acta Geod. Geophys. Hung. 46(3): 347–369

[93] E. Győri, T.J. Katona, Z. Bán, L. Tóth (2014) Methods and uncertainties in liquefaction hazard assessment for nuclear power plants. In: Proceedings of Second European Conference on Earthquake Engineering and Seismology: EAEE Sessions. Istanbul, Turkey, 2014.08.25-2014.08.29, pp. 535–546.

[94] T.J. Katona, E. Győri, Z. Bán, L. Tóth (2015) Assessment of liquefaction consequences for nuclear power plant Paks. In: 23th Conference on Structural Mechanics in Reactor Technology. Manchester, UK, 2015.08.10-2015.08.14. Paper ID 125.

[95] Z. Bán, E. Győri, T.J. Katona, L. Tóth (2015) Characterisation of liquefaction effects for beyond-design basis safety assessment of nuclear power plants. Geophysical Research Abstracts 17: Paper EGU2015-4152. (2015), European Geosciences Union General Assembly 2015. Vienna, Austria: 2015.04.12–2015.04.17.

[96] T. J. Katona (2015) Safety assessment of the liquefaction for nuclear power plants. Pollack Periodica Int. J. Eng. Inf. Sci. 10(1):39–52

[97] T. J. Katona, Z. Bán, E. Győri, L. Tóth, A. Mahler (2015) Safety assessment of nuclear power plants for liquefaction consequences. Sci. Technol. Nucl. Install. p. 11 (Paper 727291)

[98] Z. Bán, T. J. Katona, A. Mahler (2016) Comparison of empirical liquefaction potential evaluation methods. Pollack Periodica Int. J. Eng. Inf. Sci. 11(1):55–66

[99] Position paper on Periodic Safety Re-views (PSRs) taking into account the lessons learnt from the TEPCO Fukushima Dai-ichi NPP accident, Study by WENRA Reactor Harmonization Working Group, March 2013. http://www.wenra.org/media/filer_public/2013/04/05/rhwg_position_psr_2013-03_final_2.pdf

[100] General Presentation of the HERCA-WENRA Approach for a better cross-border coordination of protective actions during the early phase of a nuclear accident, HERCA, WENRA, Stockholm, 22 October 2014

[101] Safety of nuclear power plants: Design, Specific Safety Requirements, No. SSR-2/1 (Rev. 1), International Atomic Energy Agency, Vienna, 2016, ISBN 978–92–0–109315–8

[102] IAEA, 2003, NS-G-1.6, Seismic Design and Qualification for Nuclear Power Plants: Safety Guide, International Atomic Energy Agency, Vienna, 2003, ISBN 92–0–110703–X

[103] ASME Boiler and Pressure Vessel Code (BPVC), Section III: Rules for Construction of Nuclear Power Plant Components, Division 1, American Society of Mechanical Engineers/2015. ISBN: 9780791869802

[104] Standards for Design of Seismic Resistant Nuclear Power Plant NP-031-01, Gosatom-nadzor, Moscow, 2002. http://en.gosnadzor.ru/framework/nuclear/NP-031-2001.pdf

[105] ASME (2012a) STP-NU-051, Code Comparison Report for Class 1 Nuclear Power Plant Components, ASME Standards Technology, LLC, 2012. ISBN No. 978-0-7918-3419-0

[106] NRC (2006) Regulatory Guide 1.201, Guidelines for categorizing structures, systems, and components in nuclear power plants according to their safety significance, Revision 1, May 2006. http://pbadupws.nrc.gov/docs/ML0610/ML061090627.pdf

[107] EN 1998 EUROCODE 8, Design of structures for earthquake resistance, EN 1998-1 General rules, seismic actions and rules for buildings

[108] Program on Technology Innovation: Validation of CLASSI and SASSI Codes to Treat Seismic Wave Incoherence in Soil-Structure Interaction (SSI) Analysis of Nuclear Power Plant Structures. EPRI, Palo Alto, CA: 2007. 1015111.

[109] B. Jeremic, N. Tafazzolia, T. Anchetac, N. Orbovicd, A. Blahoianu (2013) Seismic behavior of NPP structures subjected to realistic 3D, inclined seismic motions, in variable layered soil/rock, on surface or embedded foundations. Nucl. Eng. Des. 265:85–94. doi:10.1016/j.nucengdes.2013.07.003

[110] D. M. Ghiocel, S. Short, G. Hardy (2009) Seismic motion incoherency effects on SSI response of nuclear islands with significant mass eccentricities and different embedment levels. In: 20th International Conference on Structural Mechanics in Reactor Technology (SMiRT 20) Espoo, Finland, August 9–14, 2009, SMiRT20-Division 5, Paper 1853

[111] P. Rangelow, W. Schütz, L. Muszynski, B. Kross (2015) Modeling effects on the response of a reactor building under incoherent seismic ground motion excitation, transactions, SMiRT-23, Manchester, United Kingdom, August 10–14, 2015, Division V, Paper ID 229

[112] A Methodology for Assessment of Nuclear Power Plant Seismic Margin (Revision 1), EPRI NP-6041-SL, Revision 1, Project 2722-23, Final Report, August 1991

[113] A. Gürpinar, A. Godoy, A. R., Johnson, J.J. Considerations for Beyond Design Basis External Hazards in NPP Safety Analysis, Transactions, SMiRT-23, Manchester, United Kingdom, August 10-14, 2015, Division IV, Paper ID 424

[114] General Requirements for Seismic Design and Qualification of CANDU Nuclear Power Plants, CSA N289.1, Canadian Standards Association, Rexdale, Ontario, Canada

[115] Appendix S of the 10 CFR Part 50, Earthquake Engineering Criteria for Nuclear Power Plants, http://www.nrc.gov/reading-rm/doc-collections/cfr/part050/part050-apps.html

[116] Prestandard and Commentary for the Seismic Rehabilitation of Buildings, Federal Emergency Management Agency FEMA 356/November 2000. http://www.conservationtech.com/FEMA-publications/FEMA356-2000.pdf

[117] EN 1998-3 EUROCODE 8: Design of structures for earthquake resistance -Part 3: Assessment and retrofitting of buildings, 15 March 2005. https://law.resource.org/pub/eu/eurocode/en.1998.3.2005.pdf

[118] Design of reactor containment systems for nuclear power plants, Safety standards series, No. NS-G-1.10, International Atomic Energy Agency, 2004, Vienna, ISBN 92–0–103604–3

[119] W. E. Vesely, F. F. Goldberg, N. H. Roberts, D. F. Haasl, Fault Tree Handbook, NUREG-0492. January 1981, U.S. Nuclear Regulatory Commission Washington, DC 20555. http://www.nrc.gov/docs/ML1007/ML100780465.pdf

[120] P. G. Prassinos, M. K. Ravindra, J. D. Savay, Recommendations to the Nuclear Regulatory Commission on Trial Guidelines for Seismic Margin Reviews of Nuclear Power Plants, Lawrence Livermore National Laboratory, NUREG/CR-4482, 1986.

[121] ASME/ANS RA-S-2008, Standard for Level 1/Large Early Release Frequency Probabilistic Risk Assessment for Nuclear Power Plant Applications, ASME, 2008, ISBN: 9780791831403

[122] J. Prochaska, P. Halada, M. Pellissetti, M. Kumar, Report 1: Guidance document on practices to model and implement SEISMIC hazards in extended PSA, Reference ASAMPSA_E, Technical report ASAMPSA_E/ WP22/ D22.2-1 2016-19, Reference IRSN PSN/RES/SAG/2016-0233, http://asampsa.eu/wp-content/uploads/2016/05/ASAMPSA_E-WP22-D22.2-3-REPORT-1-Seismic-V1.pdf

[123] R. J. Budnitz et al. (1985) An approach to the quantification of seismic margins in nuclear power plants, Lawrence Livermore National Laboratory, NUREG/CR-4334

[124] D. A. Heacock (2011) North Anna Power Station Restart Readiness, November 1st 2011, Public Meeting. http://www.nrc.gov/info-finder/reactor/na/dominion-slides-11-01-2011-meeting.pdf

[125] Virginia Electric and Power Company (Dominion) North Anna Power Station Units 1 and 2, North Anna Independent Spent Fuel Storage Installation—Summary Report of August 23. 2011, Earthquake Response and Restart Readiness Determination Plan, Virginia Electric and Power Company, Richmond, Virginia 23261, September 17, 2011. https://www.dom.com/about/stations/nuclear/north-anna/pdf/Earthquake_Summary_Report_and_Restart_Plan_091711.pdf

[126] K. Decker, Bibliography—Existing Guidance for External Hazard Modelling, Reference, ASAMPSA_E, Technical report ASAMPSA_E/WP21/D21.1/2015-09, Reference IRSN PSN-RES/SAG/2015-00082, http://asampsa.eu/wp-content/uploads/2016/06/ASAMPSA_D21_1_External_Hazards_Bibliography.pdf

[127] I. Kuzmina, A. Lyubarskiy, P. Hughes, J. Kluegel, T. Kozlik, V. Serebrjakov The Fault Sequence Analysis Method to Assist in Evaluation of the Impact of Extreme Events on NPPs, Nordic PSA Conference—Castle Meeting-2013, 10–12 April 2013, Stockholm,

Sweden https://nucleus.iaea.org/sites/gsan/lib/Lists/Papers/Attachments/3/Paper%20 S6-1%20IAEA-The%20fault%20sequence%20analysis%20method.pdf

[128] N. Siu, D. Marksberry, S. Cooper, K. Coyne, M. Stutzke (2013) PSA Technology Challenges Revealed by the Great East Japan Earthquake, PSAM Topical Conference in Light of the Fukushima Dai-Ichi Accident, Tokyo, Japan, April 15-17, 2013. http://www.nrc.gov/docs/ML1303/ML13038A203.pdf

[129] A. Lyubarskiy, I. Kuzmina, M. El-Shanawany (2011) Notes on Potential Areas for Enhancement of the PSA Methodology based on Lessons Learned from the Fukushima Accident, Presented at UK' s 2nd Probabilistic Safety Analysis / Human Factors Assessment Forum, 8-9 September 2011, The Park Royal Hotel, Warrington, UK. https://nucleus.iaea.org/sites/gsan/lib/Lists/Papers/Attachments/1/Lessons%20from%20FA%20for%20PSA_UK-Forum_Sept2011.pdf

[130] PSA of Natural External Hazards including Earthquake, Workshop proceedings, Prague, Czech Republic June 17–20, 2013, NEA/CSNI/R(2014)9. https://www.oecd-nea.org/nsd/docs/2014/csni-r2014-9.pdf

[131] PSAM2013: PSAM topical conference in Tokyo. In light of the Fukushima Dai-ichi accident; Tokyo (Japan); 14-18 Apr 2013. http://www.psam2013.org

[132] Safety Reports Series No 66, Earthquake Preparedness and Response for Nuclear Power Plants, IAEA, Vienna, 2011, ISBN 978-92-0-108810-9IAEA, 2012

[133] Pre-earthquake Planning and Immediate Nuclear Power Plant Operator Post-earthquake Actions, Regulatory Guide 1.166, U.S. NRC, 1997.

[134] Guidelines for Nuclear Plant Response to an Earthquake, Rep. EPRI-NP-6695, EPRI, Palo Alto, CA, 1989.

[135] Guidelines for Nuclear Plant Response to an Earthquake, EPRI Technical Report 3002000720, October 2013.

[136] Benchmarking for Seismic Housekeeping at Nuclear Power Plants: Compilation of Industry Practices. EPRI, Palo Alto, CA, and Seismic Qualification Utility Group (SQUG): 2008. 1018352.

[137] A. Vilimi, L. Tóth, T. J. Katona (2016) Analysis of consequences of a design basis earthquake for the region around a nuclear power plant. Pollack Periodica Int. J. Eng. Inf. Sci. 11(2):43–54. doi:10.1556/606.2016.11.2.4. www.akademiai.com

[138] T. J. Katona, L. Tóth (2013) Earthquake damage indicators for nuclear power plants. In: B. H. V. Topping, P. Iványi (editors) Proceedings of the Fourteenth International Conference on Civil, Structural and Environmental Engineering Computing, 236 p. Italy, 2013.09.03–2013.09.06. Stirlingshire: Civil-Comp Press, Kippen, Stirlingshire, 23 UK, 2013, ISBN: 978-1-905088-57-7, Paper 91.

[139] T. J. Katona, L. Tóth (2013) Damages indicators for post-earthquake condition assessment. Acta Geod. Geophys. 48:333–345. doi:10.1007/s40328-013-0021-9

[140] Restart of a Nuclear Power Plant Shut Down by a Seismic Event, Regulatory Guide 1.167, U.S. NRC 1997

[141] Criterion for determining accidence of the operating basis earthquake, EPRI NP-5930, July 1988

[142] Standardization of the Cumulative Absolute Velocity, Report No. EPRI TR-100082-T2, EPRI, Palo Alto, California, 1991.

[143] T. J. Katona (2011) Interpretation of the physical meaning of the cumulative absolute velocity. Pollack Periodica Int. J. Eng. Inf. Sci. 6:9–106

[144] T. J. Katona (2012) Modeling of fatigue-type seismic damage for nuclear power plants. Comput. Mater. Sci. 64: 22–24

3

New Insight in Liquefaction After Recent Earthquakes: Chile, New Zealand and Japan

Yolanda Alberto-Hernandez and Ikuo Towhata

Additional information is available at the end of the chapter

Abstract

Liquefaction has proved to be one of the major geotechnical issues caused by earthquakes. It is one of the most costly phenomena and has affected several cities around the world. Although the topic has been studied since the 1960s, new questions are emerging. The earthquakes of Chile in 2010, New Zealand in 2010 and 2011, and Japan in 2011 had in common not only being some of the largest earthquakes of this decade but also having a problem of extensive liquefaction. Although most seismic codes have provisions against liquefaction, there are still some misconceptions regarding the characteristics of soil susceptibility and the effect of repeated liquefaction. This chapter introduces a detailed report of the damage caused by liquefaction in the cities affected by those earthquakes and also highlights observations in liquefied areas that were unexpected. Advanced geotechnical testing was conducted and compiled to compare them with previous assessment criteria and observations. A more comprehensive framework for the evaluation of liquefaction susceptibility and countermeasures will be presented and a roadmap of future work in the area will be described.

Keywords: iquefaction, fines content, repeated liquefaction, 2010 Chile Earthquake, New Zealand Earthquake, 2011 Japan Earthquake

1. Introduction

Liquefaction is a hazard that has caused a large number of casualties and economic losses. Although the phenomenon has been observed for a long time, only relatively recently it has been acknowledged and more discoveries are being done during the latest seismic events. Dutch engineers recognized the phenomenon of strength loss and pore-pressure increment after the severe flow slides caused by vibration near a railway bridge at Weesp in 1918 [1].

Mogami and Kubo [2] reported small heavings of sand at some places as Amagasaki during an earthquake in 1951. They performed experiments on Kumiho sand and other materials, using a metal box able to move vertically and sinusoidally. They found that as acceleration increased, shearing strength decreased to almost zero which made the soil behave as a liquid and they decided to call this phenomenon liquefaction. Nevertheless, it was until the earthquakes of 1964, in Anchorage, Alaska and Niigata, Japan, that liquefaction was acknowledged as an important engineering problem. From that point, a vast research has been conducted on liquefaction. However, recent earthquakes in Chile, New Zealand and Japan have proved that there is still a gap of knowledge in this topic regarding susceptibility of silty soils, repeated liquefaction and ageing effects.

1.1. Chile Earthquake

A M_w = 8.8 earthquake hit the west coast of Maule, Chile, on February 27, 2010, at 3:34 local time. The epicentre, 335 km away from Santiago, was located offshore at 35.909°S, 72.733°W with a depth of 35 km and had a plate rupture area of about 550 by 150 km. This earthquake was the second largest in the Chilean History and was accompanied by a large number of aftershocks that continued over several months and reached magnitudes higher than six. The area affected included the cities of Santiago, Vina del Mar, Angol and Concepcion in three different regions as shown in **Figure 1**. One of the most distinctive characteristics of this earthquake was the long duration of large accelerations. In some areas, values greater than 0.05 g lasted longer than 60 s or even 120 s.

A group of Japanese experts (Architectural Institute of Japan, Japan Association of Earthquake Engineering, Japanese Geotechnical Society and the Japanese Society of Civil Engineering) along with Chilean specialists prepared a report regarding liquefaction in the affected areas of Chile [3]. Their reconnaissance showed that given the magnitude of the earthquake, several places experienced liquefaction, although given the season of the year, the groundwater table was low and only sites located near a saturation source liquefied. The soil was found to be composed of quaternary deposits in the coastal area, the general stratigraphy included alluvial deposits, strata of loose sandy silt and a sand backfill, where liquefaction usually started [4].

Several buildings, including modern structures, underwent differential settlement in the Concepcion area and some buried tanks emerged due to liquefaction. Ports, a fundamental infrastructure for the Chilean economy, were affected by liquefaction and lateral spreading. Bridges such as the Llacolen bridge, the Juan Pablo II bridge and the Bio-Bio bridge showed column shear failures and pier settlement [5]. There was also damage in dams, slopes and embankments, where cracks were observed. In tailing dams, the remaining of ore account for particles as fine as silt which increased the liquefaction potential. Mines as Curico, Veta del Agua and La Florida exhibited signs of liquefaction [4]. Liquefaction also destroyed several water distribution pipelines and large-diameter steel transmission pipelines, but high-density polyethylene (HDPE) pipes remained undamaged as reported by Duhalde [6].

On the other hand, in residential developments, it was observed that mitigation measures such as dynamic compaction were very effective against liquefaction, for instance, at the Ribera Norte Bío Bío housing project [3].

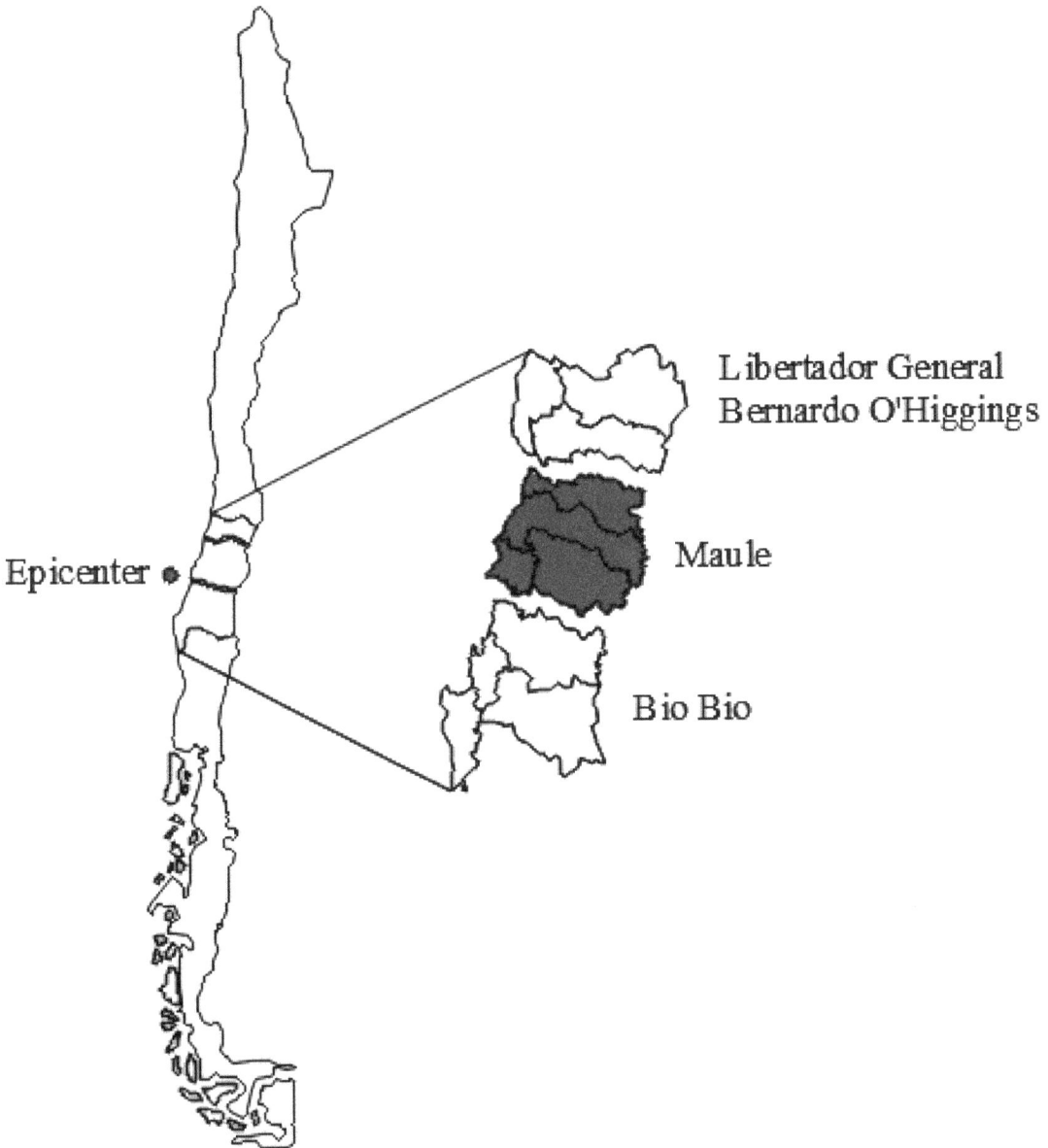

Figure 1. Areas affected and epicentre location, Chile Earthquake 2010.

1.2. New Zealand Earthquake

Two earthquakes struck the New Zealand Island on September 4, 2010, and February 22, 2011. The first one, $M_w = 7.1$, is located in the Darfield area and the second, $M_w = 6.3$, in the Christchurch area. Extensive liquefaction, affecting primarily the residential areas and pipelines, was observed (**Figure 2**). The lifelines were severely compromised and recovery was a complex task given the continuous aftershocks and their large magnitudes.

During the Darfield earthquake, there were clear signs of liquefaction in the eastern part of Christchurch near the Avon River in Avonside, Dallington, New Brighton and Bexley. At that time, the Central Business District (CBD) was not severely affected but later, during the February earthquake, the liquefaction extended to the southern part and caused even more damage leading to a 'flood' due to the large amounts of liquefied soil [7].

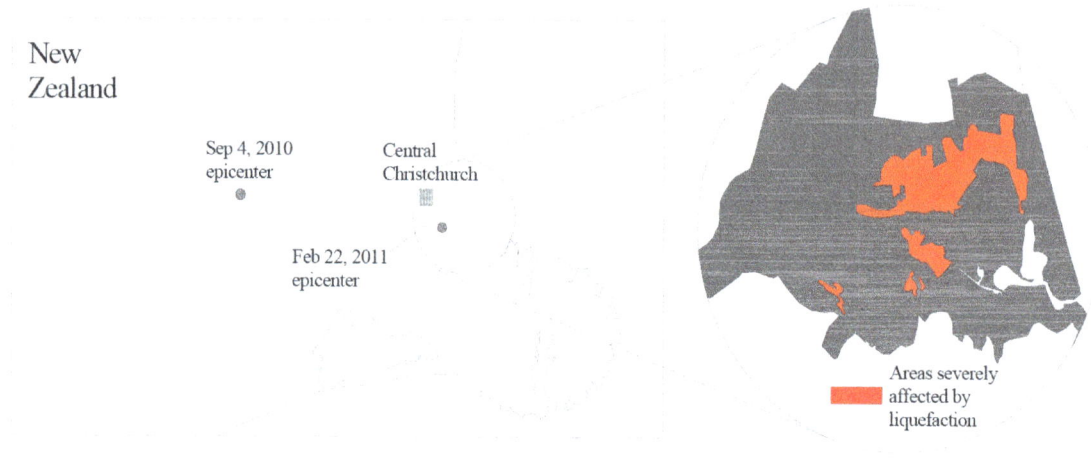

Figure 2. Areas affected and epicentre locations of the New Zealand Earthquakes.

The City of Christchurch is located on the east coast of New Zealand, and it was mainly built on reclaimed swamp. Soils in the area are clean sands around the Avon River, along with loose silts and peat in the southeast part of the CBD. Boiled sand along the Avon River contained 5–20% of silty fines and the fines were low to non-plastic [8].

Damage in the residential area was extensive in the east and northeast of the CBD, where soils are mostly clean fine to medium sands with non-plastic silt. More than 15,000 residential properties and buildings were affected particularly due to lateral spreading and differential settlement. The evaluation of the liquefaction potential of soils containing non-plastic fines is of major interest in the prevention of future damage in the city. Some studies have been conducted on the effect of fines content (FC) in the sandy soils of the surroundings in this area, finding a more contractive behaviour with the addition of fines when density measures as void ratio or relative density are used (e.g., [8, 9]).

Some other remarkable characteristics in these events are the cumulative effects of these strong earthquakes and repeated liquefaction. In different areas, it was observed that sites liquefied in previous earthquakes re-liquefied and sometimes with more intensity than the previous times.

1.3. Japan Earthquake

The 2011 off the Pacific Coast of Tohoku Earthquake, M_w = 9.0, hit the east coast of Japan on March 3 and triggered a tsunami (**Figure 3**). The disaster caused a tremendous number of casualties and economic loss and became a watershed in earthquake and risk engineering due to the combined events which also included a nuclear accident in the Fukushima plant. This event, one of the five most powerful earthquakes in the world since 1900, was followed by two aftershocks inducing additional damage of M_w = 7.4 and M_w = 7.7, 15 and 30 min after the first event.

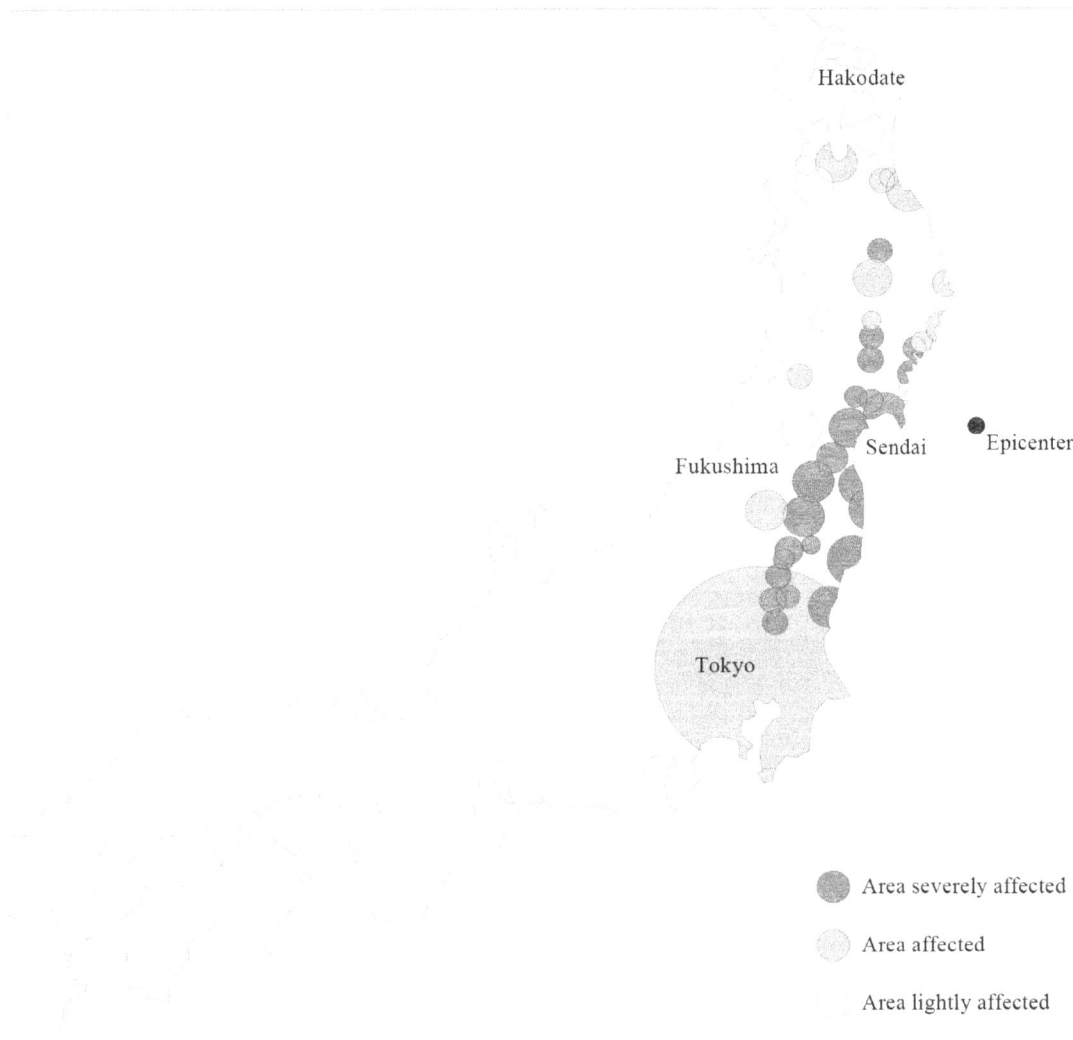

Figure 3. Areas affected and epicentre location of the 2011 Great East of Japan Earthquake.

Several reports (e.g., Refs. [10–15]) showed that liquefaction occurred along the coast of Japan, although most of the traces were erased by the tsunami in the northern part. In spite of being more than 200 km away from the earthquake fault, the Kanto area was significantly affected, especially the Tokyo Bay, where there is a large extension of reclaimed land. Severe liquefaction-induced damage was observed in Tokyo Bay, where reclamation of the coastline started around 1600, from Sumida to Yokohama and expanded to Kanagawa and Chiba Prefectures around the 1950s. These areas were affected before by other seismic events. On December 17, 1987, a $M_w = 6.7$ Earthquake hit eastern Chiba Prefecture (Chibaken Toho-oki Earthquake) causing liquefaction in many areas of the city. Therefore, some zones that experienced liquefaction in 1987 liquefied again in 2011.

Urayasu City, a zone of reclaimed land, was built in three different stages; the first reclamation was done before 1945 and later had extensions in 1948, 1968, 1971, 1975, 1980 and 2009. The reclamation work was done by the hydraulic method, consisting in dredging the shallow seabed and transporting the soil hydraulically by pipes and having the soil sedimented under water. There was more damage in Ichikawa City at the east of Urayasu, in Funabashi City and Makuhari City.

On March 2011, boiled sand was observed in the reclaimed cities of Urayasu, Ichikawa, Narashino, Odaiba, Shinonome, Tatsumi, Toyosu, Seishin, Yokohama, Kawasaki, Kizarasu and Chiba [15]. However, the most devastating effects were found on lands reclaimed after the 1970s where differential settlement and lateral displacement affected Residential areas, roads, sea walls, pipelines and other structures. Amid those areas, Urayasu City was the most affected, where more than 9000 private properties had detriment.

As stated in the previous sections, there were distinctive features in the liquefaction events; however, one of the coincidences is the presence of non-plastic fines in the sand (also considering the tailings) and the increased susceptibility to liquefaction. This topic has been debated since it is believed in the current practice that the addition of fines will decrease the liquefaction potential. Other observed phenomena were repeated liquefaction and ageing effects and will be discussed in the following sections.

2. Liquefaction of sand containing non-plastic fines

2.1. Previous cases of liquefaction of sand containing non-plastic fines

On October 17, 1989, the San Andreas fault in California ruptured over a length of approximately 45 km and generated a $M_w = 6.9$ earthquake. In the San Francisco Bay Area, there were hydraulic fills that had 3–9 m of loose, silty sand. Liquefaction-induced damage in building, infrastructure and pipelines was found mostly in the South Market area, Mission Creek and Mission District. One of the most representatives proves of liquefaction during the 1989 Loma Prieta Earthquake was the extreme damage of the Valencia Street Hotel. Most of these fills were placed during the 1900s and were composed of sand that was deposited in hydraulic suspension and allowed to settle freely [16]. Sand dredged from San Francisco Bay contained fines compared to clean Dune sand that was also present as fill material. It was reported that sand with fines had lower values of density and had more tendency to liquefy than clean sand [17].

In a study conducted by Rollings and McHood [18], the liquefaction-induced settlement in Marina District was computed and compared to the measured values. They used a correction for fines adjusting the volumetric-strain curves as pointed out by Seed et al. [19] instead of modifying the SPT N-value. Their results had a difference to the measured value of about a factor of 2, which led them to conclude that more studies should be done on defining the effect of fines on the correlations.

Another example of liquefaction on silty ground was observed during the Taiwan Earthquake. On September 21, 1999, the mountainous village of Chi-Chi was the epicentre of a $M_w = 7.6$ earthquake causing extensive liquefaction damage in foundations, embankments, riversides, retaining walls, and so on. The counties affected, Yunlin, Zhangua, Nantou and Taichung, are in Central Taiwan where soils were mostly compressible sands with large amounts of low to medium plastic fines. These soils originated from the process of weathering and abrading of shales, slates and sandstones from the central mountains; at some spots, there are layers of very loose sand susceptible to liquefaction, their fines content ranges from 10 to 50% and some of these layers are capped by thick layers of clay material [20].

Back analyses performed on the liquefaction potential of the soil showed discrepancies between the results using simplified methods and the actual observed response. For instance, [21] evaluated the methods by Refs. [22] and [23] to observe the adequacy of these procedures to be applied on the soil conditions in Central Taiwan. They found that one of the major differences is the correction factor used for fines content, while the factor of safety computed for the liquefied area was similar in both methods for fines content less than 35%. Tokimatsu and Yoshimi's correction for fines caused an overestimation and Seed's, an underestimation for the real correction in their study. Ni and Fan suggested correction of fines for the simplified methods they discussed.

Similarly, Juang et al. [20] proposed a model based on artificial neural network of limit-state data that resulted in more accuracy for considering more fines than 35%, than the method by Youd and Idriss [24]. The extensive economic loss during this earthquake also enforced the development of new methods for sampling, testing and evaluating the liquefaction potential of sands containing large amounts of fines.

2.2. Current treatment of soils containing fines

In the late 1970s, researchers as Seed and Tokimatsu developed different procedures for evaluating the liquefaction potential. Observations of liquefied sites, where it was observed that liquefaction also occurred in deposits formed by different materials as gravel and silt, were added to various databases and used for guidelines. Currently, some of the simplified procedures used worldwide are those proposed by Seed et al. [22], the Japan Road Association (1990 and 1996), [23], the Chinese Building Code (1989) or the Arias intensity method [25].

Case studies [19, 26] have shown that the existence of fines in sands increases the liquefaction resistance at the same level of standard penetration test, N-value.

The first approach to liquefaction of sands containing fines was taken on by Wang [27] who compiled a series of liquefaction events in different soils to estimate the liquefaction potential of silty soils according to its fines content, FC, plasticity index, PI, water content, wc, and liquid limit, LL. Later, Seed et al. [28] summarized Wang's findings into the three following conditions for soils vulnerable to liquefaction:

FC < 15% (per cent finer than 0.005 mm)

LL < 35%

wc > 0.9 LL

Seed et al. [28] compiled data of silty sand from liquefied sites and added them to a chart of cyclic stress ratio and modified penetration resistance. They concluded that the boundary between liquefiable and non-liquefiable soils is significantly higher for silty sands than that for clean sands. After the liquefaction events in Adapazarı during the Kocaeli Earthquake of 1999, Bray and Sancio [29] studied the limits proposed by the so-called Chinese criteria. They concluded that the use of FC for separating liquefiable from non-liquefiable soils should be avoided, redefined the relation between water content and liquid limit as wc >0.85 LL and stated that plasticity index, PI, was a good index of liquefaction susceptibility since soils with PI < 12 can liquefy.

The presence of fines during liquefaction has caused divergent conclusions regarding its effects. While field test data of sites with fines have been added in charts for design (e.g., [19, 30]), there is no clear differentiation between plastic and non-plastic fines.

Robertson and Campanella [30] in their studies on cone penetration tests found that silty sands and silts cause a decrease in penetration resistance. According to this, soils with fines at the same penetration resistance have greater liquefaction resistance than clean sand.

Given the advantages of testing soil in controlled environments, laboratory testing has been a very recurrent choice when dealing with the influence of fines on the undrained behaviour of sands. While testing, it becomes necessary to keep one parameter constant to observe the effect of the variation of others. Some of the most common parameters to keep constant during comparison are overall void ratio or simply void ratio, e, and relative density, D_r, which are good measures of particle contact. When testing clean sand, it is easy to compare results while keeping constant both of them, which has made these parameters quite useful and widespread. For that reason, in experiments with silty sand, most researchers have employed them. However, there are different issues when testing sand containing fines, which have encouraged researchers to not only understand the limitations of void ratio and relative density but also develop different parameters for comparison, as explained in the following section.

Although gradation and mineralogy of sand as well as the amount of fines tested are key factors, the difference in the results obtained by several researchers might be explained by considering the concept of void ratio. While sand has no fines, voids are only occupied by water (in a saturated soil) and void ratio is an index of particle contact and force transmission. As a small amount of fines is added to the sand matrix, voids are occupied by water and fines, reducing global void ratio although there is no contribution of the fines to the intergranular force transmission. If fine content increases, it reaches a threshold point B (**Figure 4**) when fines fill all the voids. From such a point, fines start gradually influencing the mechanical behaviour, until sand grains are fully surrounded by them and do not make contact with each other anymore; then the force is totally supported by fines. It can be deduced that the concept of void ratio as an index of particle contact is not valid after the threshold point. In this regard, variations in void ratio have been used to be representative of the behaviour of silty sand, such as the intergranular contact index void ratio [31] and the equivalent void ratio [32] both shown in **Figure 5**. These parameters seem to solve the disjunctives concerning real particle contact. However, there are still uncertainties regarding the values that must be used when fine content is very high or regarding the parameter that reflects the fraction of fines participating in the force structure of the solid skeleton (b). Some researchers as Rahman and Lo [33] have shown formulas for estimating (b), but they require different assumptions and an iterative process.

Nevertheless, since it is important to be able to compare soils with different fines content at their natural state in ground, in this paper another standpoint is taken.

The use of density measures for comparison has made the laboratory research on liquefaction of silty sand ambiguous, given the restrictions of each parameter and the impossibility to keep them all constant at the same time.

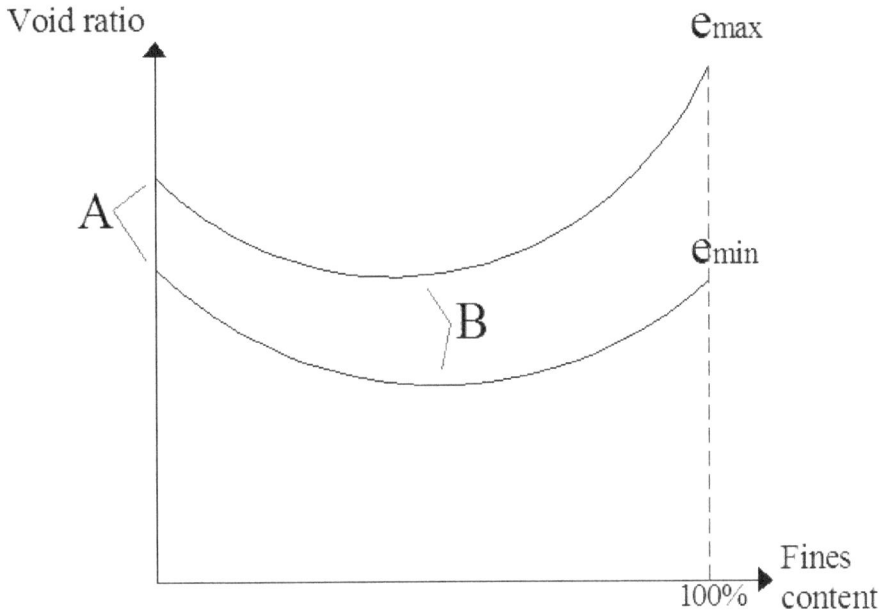

Figure 4. Theoretical variation of void ratio in binary packings with fines. After Ref. [34].

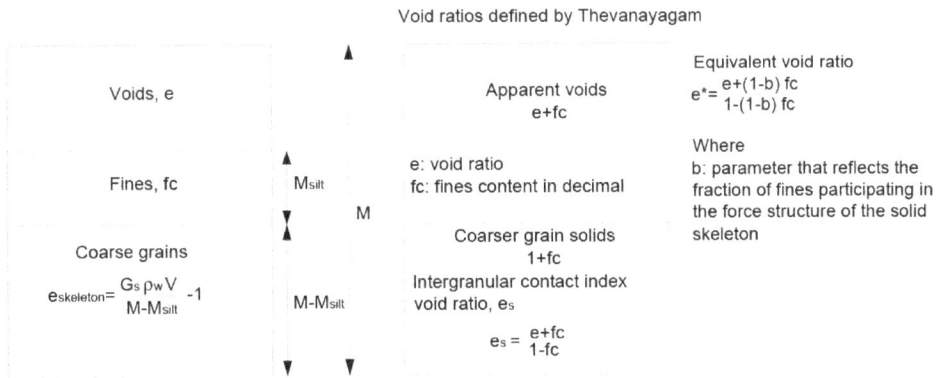

M: mass
M_{silt}: mass of silt
G_s: specific density
V: volume
ρ_w: density of water

Figure 5. Definition of sand skeleton void ratio, intergranular contact index void ratio and equivalent void ratio. *M:* mass; M_{silt}: mass of silt; G_s: specific density; *V:* volume; ρ_w: density of water.

Lee and Fitton [34, 35] performed tests on alluvial sand and gravel deposits at El Monte, Los Angeles, CA. Grain particles were composed of quartz, feldspar and dark minerals; fines varied from 0 to 90%, the fines being a mixture of silt and clay. Samples were isotropically consolidated to 15 psi (100 kPa) and pulsating-loading triaxial tests were conducted at relative densities of 50 and 75%. They found that very fine sands and silty sands showed the weakest response.

Iwasaki and Tatsuoka [36] performed tests with a resonant column apparatus with a hollow cylindrical specimen of sands with different gradations and fine content from 0 to 33%. While keeping constant void ratio, it was seen that sands other than clean sands had smaller shear moduli.

Shen et al. [37] conducted one of the first researches carried out on the effect of fines in the liquefaction potential. In their tests, they used a triaxial machine that allows for cone penetration tests on specimens with the same stress conditions as those of the static and cyclic triaxial tests. They used Ottawa sand and clayey silt with $PI = 11$ and observed that, at the same sand skeleton void ratio, fines increase the liquefaction resistance.

These primal experiments on sand with fines provided some insight on the expected influence of fines, according to the parameter used for comparison for further research. For instance, when keeping constant void ratio it has been found that liquefaction resistance decreases as fines rise (e.g., [38–42]). If relative density is held constant used for comparison, liquefaction resistance grows with the addition of fines (e.g., [40, 42–44]). Some researchers as Kuerbis [43] found the sand skeleton void ratio, which assumes that the volume occupied by fines is part of the volume of voids, to be a more appropriate parameter because it seemed to be independent of fines content; yet, Polito and Martin [45] identified a growth in liquefaction resistance with fine content for Yatesville sand when maintaining constant sand skeleton void ratio.

Liquefaction resistance in silty sand has demanded the attention of many researchers throughout the years. When researchers compared the same parameter, they found similar conclusions. It is important to note that most researchers have focused only on fines content below 30%, which is usually the limit for using parameters as void ratio, relative density, sand skeleton void ratio or even equivalent void ratio.

2.3. Hollow torsional shear tests conducted on sand containing non-plastic silt

Torsional shear tests were conducted on boiled silty sand collected from Urayasu City after the 2011 Tohoku Earthquake in Japan, hereinafter called Urayasu sand. A typical grain size distribution is shown in **Figure 6**.

The variation in minimum and maximum void ratios with fines content is shown in **Figure 7**, and it can be seen that there is a "V-shape" in these curves, as pointed out by Lade et al. [34]. The lower minimum and maximum void ratios indicate the threshold value where the voids in the sand matrix are completely filled with fines, having the lower resistance condition at the bottom part of the curve (fine content between 30 and 40% for the maximum void ratio).

As stated before, as more fines are added, minimum and maximum void ratios vary, making it difficult to select void ratio or relative density as constant parameters for the evaluation of soil behaviour, since all parameters cannot be kept constant at the same time. To avoid this concern, constant energy for sample preparation is used in this study as the comparison parameter for several fines contents to offer a different perspective in this matter. As pointed out by Lade and Yamamuro [44], any depositional process will produce different densities depending on the gradation of soil in nature, hence the use of the same compaction energy for sample preparation is by some means the reproduction of the same natural environment, although in a very naive way.

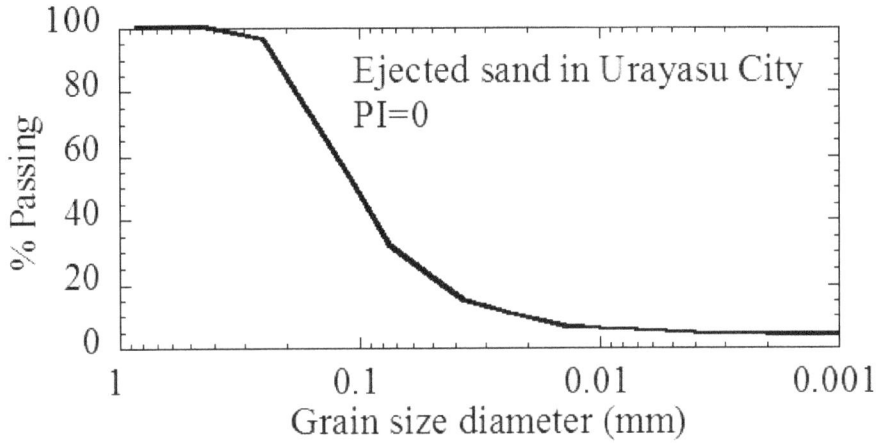

Figure 6. Grain size diameter of Urayasu sand.

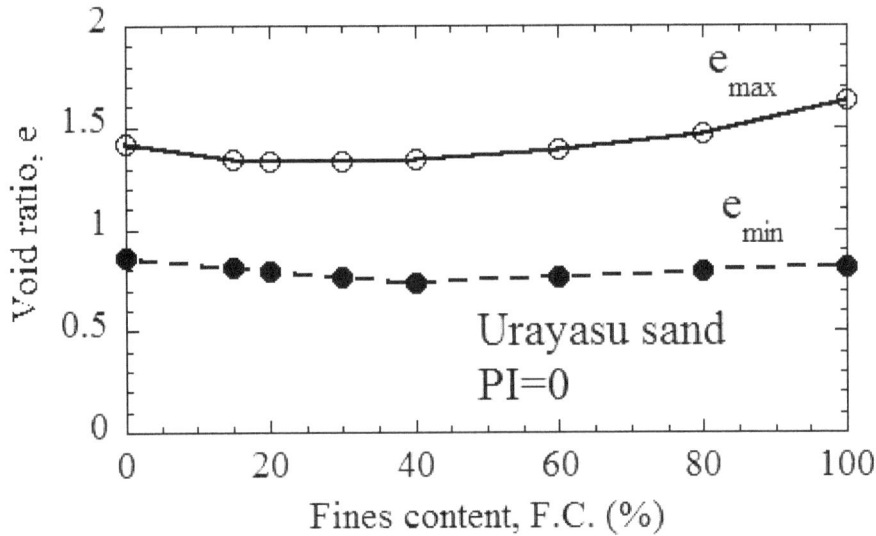

Figure 7. Minimum and maximum void ratio of Urayasu sand.

The device selected to carry out these tests was a hollow cylindrical torsional shear device that can subject a 190-mm height specimen (internal diameter of 60 mm and external diameter of 100 mm) to a combination of axial and torsional stresses, in addition to the fluid stresses inside and outside the cylindrical surfaces. Sand was sieved to separate fines from sand grains. After washing and drying, both sand and fines were thoroughly mixed together, varying the amount of fines from 0 to 80%. Dry pluviation was chosen for practical purposes since it does not overestimate the cyclic resistance [46], and some tests by Huang et al. [39] proved that dry deposition samples with fines content up to 30% shows uniformity. This procedure consists of pouring dry sand with a funnel with a fixed inner diameter and a constant height of fall, in this case 5 cm (AP-5 cm). The funnel should be turned slowly around the sample, first clockwise and then counter-clockwise to allow for even distribution of the grains; this practice was repeated keeping the same height of fall at all times, to use this energy of compaction as the parameter of comparison.

Saturation of Urayasu sand was completed using the double vacuum method described by Ampadu and Tatsuoka [47]. While keeping the specimen at a constant effective stress of 20 kPa, vacuum was incrementally applied to the inner and outer cells, as well as the interior of the specimen reaching -70 and -90 kPa, respectively. After 1 or 2 h of vacuuming, depending on the amount of fines and density, deaired water was flushed through the specimen allowing for full saturation. According to the fines content, this process could take up to 4 h. After this, backpressure saturation was used to dissolve any air remaining in the voids. The degree of saturation was measured with Skempton's B-value and all samples reached values of $B \geq 0.96$. Once satisfactory B-values were achieved, sand was isotropically consolidated to an effective confining stress σ'_{c0} of 100 kPa. During this stage, volumetric and axial strains were measured. After consolidation, cyclic shear tests were conducted with a strain rate of 0.12%/min. Several cyclic stress ratios (τ/σ'_{c0}) were chosen for samples with fines content varying from 0 to 80%, in order to define liquefaction curves.

3. Volumetric strain during consolidation

One important factor to understand the effect of different fines contents is the amount of volumetric strain during consolidation. Since the energy for preparation is the same, it is possible to compare the volumetric strain during consolidation for all samples (**Figure 8**). The graphs show that there are three groups of behaviour regarding the fines content; for AP-5 cm, the 0–20% samples seem to increase their volumetric strain as the fine content increases, while the 30–40% seems to decrease the volumetric strain as the fine content increases, as well as the 60–80% group. However, in this group, due to the amount of silt, the volumetric strain is larger.

Another key factor is the coefficient of volume compressibility, m_v, which is most likely inversely proportional to the strength of soil if Young modulus is also proportional to strength. In such case, m_v can be a laboratory parameter used for evaluating strength in the field. Considering the values in **Figure 8**, the coefficient of volume compressibility was computed, and it was found that from 0 to 20%, the value becomes larger, from 30 to 40%, m_v decreases, and it also decreases from 60 to 80%, although with larger values than the previous group.

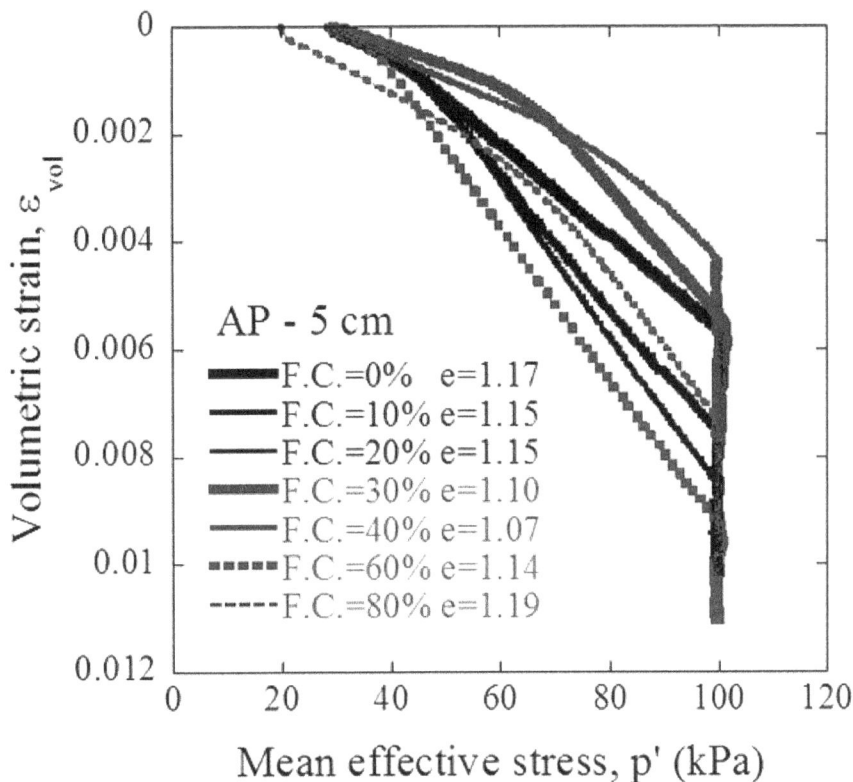

Figure 8. Volumetric strain versus mean effective stress.

4. Stress-strain curves and stress paths

Once the test programme was completed, stress-strain curves and stress paths were plotted. Results showed that there are three different behaviours, according to their relation to the threshold fines content. Below the limiting fines content, from 0 to 20% there is response dominated by the sand grains, from 30 to 50% there is a transition stage between sand and fines behaviour, then, above the threshold value, from 60 to 80%, the behaviour seems to be dominated by the contacts along the fines. Outcomes are described considering this perspective. **Figure 9** shows a comparison of the final cycle of liquefaction in which the strain amplitude is in the range of -9 to 10% for three samples with 0, 30 and 80% fines content. It can be seen that the reduction in the tangent shear modulus seems to be smaller as the fines content increases.

The corresponding stress paths are depicted in **Figure 10**, and it is noted that the 80% curve does not reach the zero-effective stress point as the other curves. It is also seen that there are very small differences between the 30 and 80% samples formed at different densities.

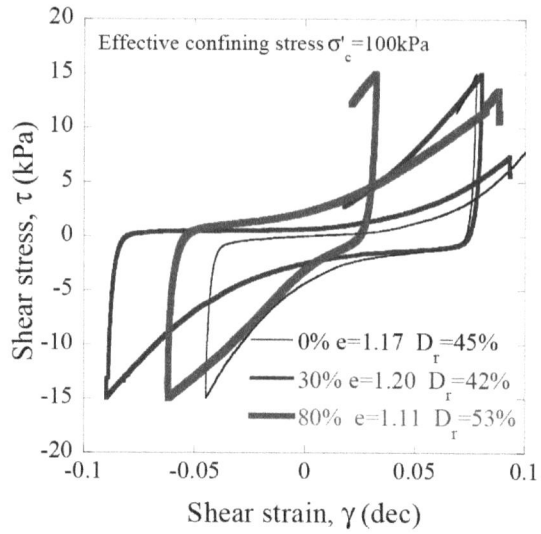

Figure 9. Stress-strain comparison of samples containing 0, 30 and 80% of fines.

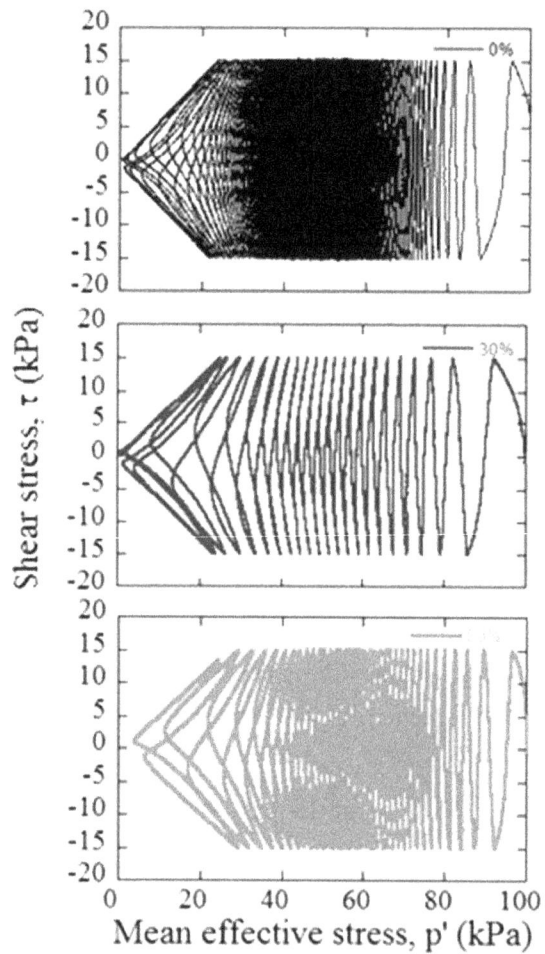

Figure 10. Stress-path comparison of samples containing 0, 30 and 80% of fines.

5. Results of cyclic shear tests

Two criteria were used for defining liquefaction, the generation of total excess pore-pressure ratio, $r_u = 1$ and the 5% double amplitude of shear strain. Given the nature of loose samples, they yielded similar results for AP-5 cm. The curves shown in **Figure 11** were constructed using the criterion of 5% double amplitude of shear strain for liquefaction.

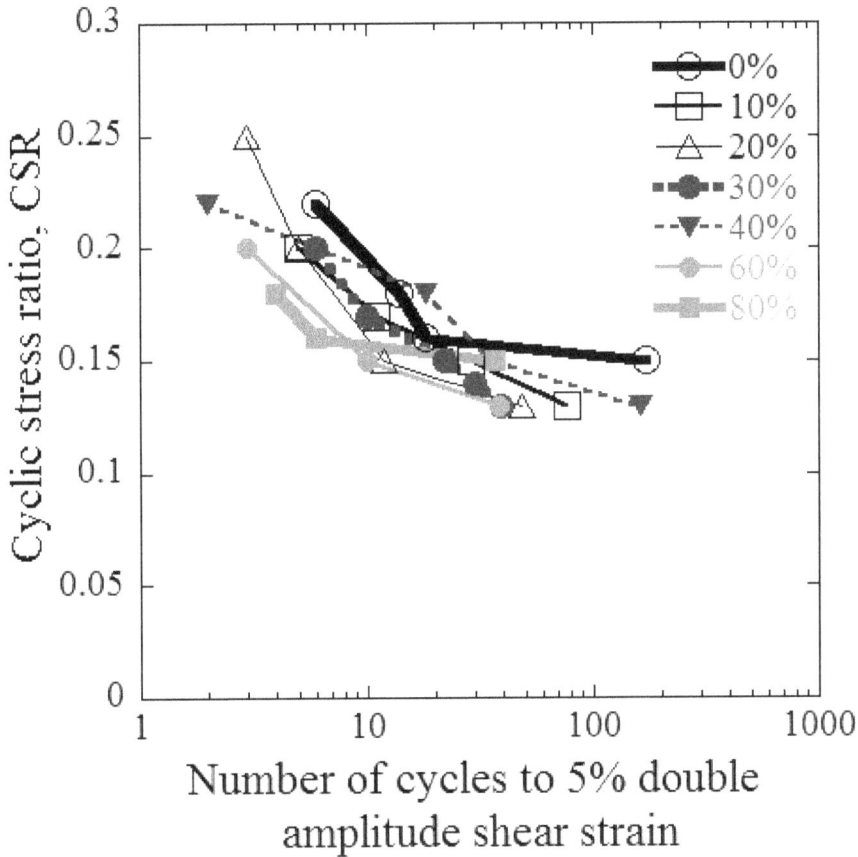

Figure 11. Liquefaction curves for samples containing 0–80% fines content.

From the series of tests, it can be seen that there are three noteworthy groups within the fines content, which are sand-like behaviour, intermediate behaviour and clay-like behaviour. It is important to remark that the clay-behaviour in this paper does not refer to clayey material but to clay-size material. In the same way, sand-like behaviour refers to the typical response of granular materials. It can be observed that the clean sand specimen has larger resistance than the samples that have fines.

There is a distinction from 0 to 20% of fines, where the resistance drops as the fines content increases; from 30 to 40%, where there is the threshold of the maximum amount of fines that can be fit in the sand matrix voids, there is a rise in resistance as fines content augment. Finally, there are the high contents of silt, from 60 to 80%, where the sand loses contact between grains and each grain is surrounded by silt which controls the response of soil. In this group, there is an overall reduction in the resistance, compared to the first two groups, but the larger the fines content, the larger the resistance. The more resistant samples within their respective ranges of fines content are the 0, 40 and 80% specimens, which might indicate that arrange of fines results in the best resistance, independently of the amount of fines content.

6. Repeated liquefaction

There were several examples of repeated liquefaction in New Zealand and Japan. The large amount of sand ejected in the residential area of Kaiapoi in New Zealand showed that repeated liquefaction represented a significant issue. Similarly, liquefaction spots identified during the Tohoku Earthquake had liquefied more than three times during previous seismic events. It could be expected that liquefaction will actually increase the relative density of soil; however, that increment could be about 10%, not large enough to reduce the liquefaction potential [14]. The major problem with repeated liquefaction is the disruption caused for reconstruction after earthquakes, especially underground lifelines and foundations, given the cumulative damage that increases the impact.

Wakamatsu [48] presented a series of borehole data in Chiba Prefecture after the 1987 earthquake and compared it to the damage observed in 2011. They found that similar damage patterns and the sites where repeated liquefaction was observed consisted of alluvial deposits or artificially filled areas.

7. Ageing effect

The ageing mechanism has brought up several questions given that observations seem to disagree with existing procedures to estimate liquefaction resistance. Aged sand having similar SPT N-values to those of younger sand has exhibited greater liquefaction resistance. During the 2011 Great East Japan Earthquake, this phenomenon was observed again in the coast of Tokyo, where only soils reclaimed after 1960 liquefied. Kokusho et al. [49] conducted a series of experiments on sands containing different fines contents and found a correlation between cone resistance and liquefaction strength independent from relative density and fines contents, which is contrary to the current practice. Their specimens of Futtsu sand with cement to simulate geological age showed higher liquefaction strength compared to specimens without cement at the same cone resistance. Towhata et al. [50] compiled data from liquefied and unliquefied sites to compare the factor of safety for liquefaction, F_L, and the time of land reclamation. They observed that some soils with values of F_L lower than one but older than 1960 did not liquefy.

8. Future trends and conclusions

This chapter introduced the liquefaction-induced damage reported during recent earthquakes in Chile, New Zealand and Japan. One of the most significant issues was the liquefaction potential of sands containing non-plastic fines. The current practice indicates that sand containing fines is generally more resistant to liquefaction than clean sand; however, the observations in tailing mines, reclaimed areas and alluvial deposits showed that the amount of silt had a significant influence.

The authors conducted a review of previous experimental research on liquefaction potential of silty sands. Torsional shear tests were conducted on silty sand from Urayasu City, varying the fines content from 0 to 80% to consider all ranges of behaviour. Three different responses were found and their characteristics were evaluated in terms of excess pore-pressure build-up, shear strain and cyclic resistance ratio. Keeping the same energy for sample preparation gives some useful insight into the behaviour of silty sand. It was found, by using this criterion, that the resistance of clean sand is always greater than that of sand mixed with fines. However, the behaviour of the silty sand depends more on their relation to the limiting fines content. When it is below this value, liquefaction resistance decreases with increasing fines content, around this value liquefaction resistance increases, and for high values of fines content, soil behaviour is dominated by the fines and liquefaction resistance increases as less sand grains are immersed into the sample.

Regarding repeated liquefaction, during the events in New Zealand and the earthquakes in Japan of 1987 and 2011, it was observed that the risk of liquefaction does not decrease after the first event given that soil does not gain enough resistance in the densification process.

As for the ageing effect, the mechanism was observed during these earthquakes, as well. Researchers conducted experiments to prove the benefit of geological ageing and they found that even at similar penetration resistances, older soils exhibited larger liquefaction resistance; this is of vital importance for areas of young deposits and mapping of liquefaction risk that might underestimate the resistance of aged deposits.

Author details

Yolanda Alberto-Hernandez* and Ikuo Towhata

*Address all correspondence to: yalbertoh@gmail.com

University of Tokyo, Japan

References

[1] M. G. Jefferies and K. Been, *Soil liquefaction. A critical state approach*. Taylor & Francis, NY, United States, 2006.

[2] T. Mogami and K. Kubo, "The behaviour of soil during vibration," in *Proceedings of the Third International Conference of Soil Mechanics and Foundation Engineering, Switzerland 16th-17th August, 1953, Volume 1*, 1953, pp. 152–155.

[3] R. Verdugo, F. Villalobos, S. Yasuda, K. Konagai, T. Sugano, M. Okamura, T. Tobita, and A. Torres, "Description and analysis of geotechnical aspects associated to the 2010

Chile earthquake Descripción y análisis de aspectos geotécnicos asociados al terremoto de Chile del 2010," *Obras y Proyectos (Works and Projects Magazine), vol. 8,* pp. 27–33, 2010.

[4] S. Yasuda, R. Verdugo, K. Konagai, T. Sugano, F. Villalobos, M. Okamura, T. Tobita, A. Torres, I. Towhata, "Case History. Geotechnical damage caused by the 2010 Maule, Chile earthquake", ISSMGE Bulletin, vol. 4, no. 2, pp. 16–27, 2010.

[5] J. U. B. Moehle, R. U. C. Riddell, and R. U. C. Boroschek, "8.8 Chile Earthquake of February 27, 2010," *EERI Special Earthquake Report,* vol. 10, no. June, pp. 1–20, 2010.

[6] J. Duhalde, "27/F Chile. Lecciones de la catástrofe. 27/F Chile. Lessons of the disaster," *Essbio Special Report.* pp. 61, 2010.

[7] E. Smyrou, P. Tasiopoulou, I. Bal, G. Gazetas, and E. Vintzileou, "Structural and geotechnical aspects of the Christchurch (2011) and Darfield (2010) earthquakes in New Zealand," in *Seventh National Conference on Earthquake Engineering, 30 May-3 June, Istanbul, Turkey,* no. June, pp. 1–12, 2011.

[8] S. Rees, "Effects of fines on the undrained behaviour of Christchurch sandy soils," Doctoral thesis. University of Canterbury, New Zealand, 2010.

[9] M. Arefi, "Ground response evaluation for seismic hazard assessment," Doctoral thesis. University of Canterbury, 2013.

[10] R. Boulanger, "Liquefaction in the 2011 Great East Japan Earthquake: Lessons for US Practice," in *Proceedings of the International Symposium on Engineering Lessons Learned from the 2011 Great East Japan Earthquake, March 1–4,* pp. 655–664, 2012.

[11] K. Ishihara, "Liquefaction in Tokyo Bay and Kanto Region in the 2011 Great East Japan Earthquake," *Proceedings of the International Symposium on Engineering Lessons Learned from the 2011 Great East Japan Earthquake, Tokyo, Japan,* vol. March 1–4, pp. 63–81, 2011.

[12] K. Konagai, T. Asakura, S. Suyama, H. Kyokawa, T. Kiyota, C. Eto, and K. Shibuya, "Soil subsidence map of the Tokyo Bay area liquefied in the March 11th Great East Japan earthquake," in *Proceedings of the International Symposium on Engineering Lessons Learned from the 2011 Great East Japan Earthquake, March 1–4, 2012, Tokyo, Japan,* vol. 2, pp. 855–864, 2012.

[13] W. Lee, K. Ishihara, and C. Chen, "Liquefaction of silty sand-preliminary studies from recent Taiwan, New Zealand and Japan Earthquakes," in *Proceedings of the International Symposium on Engineering Lessons Learned from the 2011 Great East Japan Earthquake, March 1-4, 2012, Tokyo, Japan,* vol. 2011, no. Figure 2, pp. 747–758, 2012.

[14] I. Towhata, S. Goto, Y. Taguchi, and S. Aoyama, "Liquefaction Consequences and Learned Lessons During the 2011 Mw = 9 Gigantic Earthquake," *Indian Geotechnical Journal,* Vol. 43, Number 2, pp. 116–126, 2013.

[15] S. Yasuda, K. Harada, K. Ishikawa, and Y. Kanemaru, "Characteristics of liquefaction in Tokyo Bay area by the 2011 Great East Japan Earthquake," *Soils and Foundations,* vol. 52, no. 5, pp. 793–810, 2012.

[16] T. Holzer, "The Loma Prieta California Earthquake of October 17 1989 – Liquefaction," USGS, Professional Paper 1551-B, 1998.

[17] G. W. Clough, J. R. Martin, and J. L. Chameau, "The geotechnical aspects," in *Practical Lessons from the Loma Prieta Earthquake*, National Academia Press, pp. 29–63, 1994.

[18] K. M. Rollings and M. D. McHood, "Comparison of computed and measured liquefaction-induced settlements in the Marina District, San Francisco," in *The Loma Prieta California Earthquake of October 17, 1989 – Liquefaction. USGS*, pp. B223–B239, 1998.

[19] H. Seed, K. Tokimatsu, L. Harder, and R. Chung, "The influence of SPT procedures in soil liquefaction resistance evaluations," in *University of California, Berkeley. Report No UCB/EERC-84/15. October 1984*, p. 59, 1984.

[20] C. H. Juang, S. H. Yang, H. Yuan, and S. Y. Fang, "Liquefaction in the Chi-Chi earthquake-effect of fines and capping non-liquefiable layers," *Soils and Foundations*, vol. 45, no. 6, pp. 89–101, 2005.

[21] S. Ni and E. Fan, "Fines content effects on liquefaction potential evaluation for sites liquefied during Chi-chi earthquake, 1999," in *13th World Conference on Earthquake Engineering, Vancouver, B.C., Canada*, no. 2521, pp. 1–15, 2004.

[22] B. H. B. Seed, K. Tokimatsu, L. F. Harder, and R. M. Chung, "Influence of SPT procedures in soil liquefaction resistance evaluations," *Journal of Geotechnical Engineering Engineering*, vol. 111, no. 12, pp. 1425–1445, 1985.

[23] K. Tokimatsu and Y. Yoshimi, "Empirical correlation of soil liquefaction based on SPT N-value and fines content," *Soils and Foundations*, vol. 23, no. 4, pp. 56–74, 1983.

[24] T. L. Youd and I. M. Idriss, "Liquefaction Resistance of Soils: Summary Report from the 1996 NCEER and 1998 NCEER/NSF Workshops on Evaluation of Liquefaction Resistance of Soils," *Journal of Geotechnical and Geoenvironmental Engineering*, vol. 127, no. 4, pp. 297–313, 2001.

[25] R. Kayen and J. Mitchell, "Assessment of liquefaction potential during earthquakes by Arias intensity," *Journal of Geotechnical and Geoenvironmental Engineering*, vol. 123, no. 12, pp. 1162–1174, 1997.

[26] K. Tokimatsu and Y. Yoshimi, "Criteria of soil liquefaction with SPT and fines content," in *VIII WCEE, San Francisco*, p. 8, 1984.

[27] W. Wang, "Some findings in soil liquefaction," *Chinese Journal of Geotechnical Engineering*, vol. 2, no. 3, pp. 55–63, 1979.

[28] H. B. Seed, I. M. Idriss, and I. Arango, "Evaluation of liquefaction potential using field performance data," *Journal of Geotechnical Engineering*, vol. 109, no. 3, pp. 458–482, 1983.

[29] J. Bray and R. Sancio, "Assessment of the liquefaction susceptibility of fine-grained soils," *Journal of Geotechnical and Geoenvironmental Engineering*, vol. 132, no. 9, pp. 1165–1177, 2006.

[30] P. Robertson and R. Campanella, "Liquefaction potential of sands using the CPT," *Journal of Geotechnical Engineering*, vol. 111, no. 3, pp. 384–403, 1985.

[31] S. Thevanayagam, "Effect of fines and confining stress on undrained shear strength of silty sands," *Journal of Geotechnical and Geoenvironmental Engineering*, vol. 124, no. 6, pp. 479–491, 1998.

[32] S. Thevanayagam, T. Shenthan, S. Mohan, and J. Liang, "Undrained fragility of clean sands, silty sands, and sandy silts," *Journal of Geotechnical and Geoenvironmental Engineering*, vol. 128, no. 10, pp. 849–859, 2002.

[33] M. M. Rahman and S. R. Lo, "The prediction of equivalent granular steady state line of loose sand with fines," *Geomechanics and Geoengineering*, vol. 3, no. 3, pp. 179–190, 2008.

[34] P. V. Lade, C. D. J. Liggio, and J. A. Yamamuro, "Effects of non-plastic fines on minimum and maximum void ratios of sand," *Geotechnical Testing Journal*, vol. 21, no. 4, pp. 336–347, 1998.

[35] K. Lee and J. Fitton, "Factors affecting the cyclic loading strength of soil," in *Vibration effects of earthquakes on soils and foundations, ASTM, STP 450*, pp. 71–95, 1969.

[36] T. Iwasaki and F. Tatsuoka, "Effects of grain size and grading on dynamic shear moduli of sands," *Soils and Foundations*, vol. 17, no. 3, pp. 19–35, 1977.

[37] C. K. Shen, J. L. Vrymoed, and C. K. Uyeno, "The effect of fines on liquefaction of sands," in *Proceedings of IX International Conference on Soil Mechanics and Foundation Engineering, Tokyo, Vol. 2*, pp. 381–385, 1977.

[38] F. Amini and G. Z. Qi, "Liquefaction testing of stratified silty sands," *Journal of Geotechnical and Geoenvironmental Engineering*, vol. 126, no. 3, pp. 208–217, 2000.

[39] Y.-T. Huang, A.-B. Huang, Y.-C. Kuo, and M.-D. Tsai, "A laboratory study on the undrained strength of a silty sand from Central Western Taiwan," *Soil Dynamics and Earthquake Engineering*, vol. 24, no. 9–10, pp. 733–743, 2004.

[40] M. Belkhatir, A. Arab, N. Della, H. Missoum, and T. Schanz, "Influence of inter-granular void ratio on monotonic and cyclic undrained shear response of sandy soils," *Comptes Rendus Mecanique*, vol. 338, no. 5, pp. 290–303, 2010.

[41] J. Carraro, P. Bandini, and R. Salgado, "Liquefaction resistance of clean and nonplastic silty sands based on cone penetration resistance," *Journal of Geotechnical and Geoenvironmental Engineering*, vol. 129, no. 11, pp. 965–976, 2003.

[42] J. Carraro, M. Prezzi, and R. Salgado, "Shear strength and stiffness of sands containing plastic or nonplastic fines," *Journal of Geotechnical and Geoenvironmental Engineering*, vol. 135, no. 9, pp. 1167–1178, 2009.

[43] R. Kuerbis, "The effect of gradation and fines content on the undrained loading response of sand," The University of British Columbia, Master Thesis, Canada, 1989.

[44] P. V. Lade and J. A. Yamamuro, "Effects of nonplastic fines on static liquefaction of sands," *Canadian Geotechnical Journal*, vol. 34, no. 6, pp. 918–928, 1997.

[45] C. Polito and J. R. Martin, "Effects of nonplastic fines on the liquefaction resistance of sands," *Journal of Geotechnical and Geoenvironmental Engineering*, vol. 127, no. 5, pp. 408–415, 2001.

[46] J. P. Mulilis, K. Arulanandan, J. K. Mitchell, C. K. Chan, and H. B. Seed, "Effects of Sample Preparation on Sand Liquefaction," *Journal of the Geotechnical Engineering Division*, vol. 103, no. 2, pp. 91–108, 1977.

[47] S. Ampadu and F. Tatsuoka, "Effect of setting method on the behaviour of clays in tri-axial compression from saturation to undrained shear," *Soils and Foundations*, vol. 33, no. 2, pp. 14–34, 1993.

[48] K. Wakamatsu, "Recurrent liquefaction induced by the 2011 Great East Japan earth-quake compared with the 1987 earthquake," *Proceedings of the International Symposium on Engineering Lessons Learned from the 2011 Great East Japan Earthquake, March 1–4, 2012, Tokyo, Japan*, pp. 675–686, 2012.

[49] T. Kokusho, Y. Nagao, F. Ito, and T. Fukuyama, "Sand liquefaction observed during recent earthquake and basic laboratory studies on aging effect," in *Earthquake Geotechnical Engineering Design*, Springer, pp. 75–92, 2014.

[50] I. Towhata, S. Goto, Y. Taguchi, and S. Aoyama, "Unsolved engineering problems after 2011 gigantic earthquake in Japan," in *Conference of Australian Earthquake Engineering Society, Gold Coast, Queensland*, pp. 1–11, 2012.

4

The *b*-Value Analysis of Aftershocks 170 Days After the 23 October 2011 Van Earthquake (*M* w, 7.1) of the Lake Van Basin, Eastern Anatolia: A New Perspective on the Seismic Radiation and Deformation Characteristics

Mustafa Toker

Additional information is available at the end of the chapter

Abstract

In this study, we analyzed the seismic radiation and deformation characteristics of the 2011 Van earthquake during aftershock events with the support of estimated dynamic parameters (seismic *b*-value and radiation efficiency, ηR), 3D crustal cross sections of aftershock hypocenters, and deformation styles of Lake Van basin. The resulted variation in the *b*-value exhibits two dramatic changes in the *b*-value: one ($b > 1$) during the first 100 days of the mainshock and the other ($b < 1$) in the last 70 days of the mainshock. The constant b ($b = 1$) indicates a seismically active time interval and transitional variation in the *b*-value from high to low. The estimated *b*-value ($b > 1$) reveals that the aftershock sequence comprised a large number of the small and same-sized events of the Van mainshock due to the extreme material heterogeneity within the rupture zone. This indicates a general decrease in shear stress and increasing complexity in the focal area. The small value ($\eta R << 1$) of ηR implies that the amount of energy mechanically dissipated during the Van rupture process is large. This reveals that the microscopic breakdown process dominates the rupture dynamics and the whole Lake Van basin. The 3D crustal images of hypocenters suggest that the Van event originated in a strongly heterogeneous fractured setting with the aseismic sedimentary section of Lake Van. The high *b*-value combined with the low radiation efficiency (ηR) shows a strongly faulted-fractured sediment-rock formation filled with gas-fluid. This suggests that the seismic energy is intermittently released in the discrete form of aftershock events which is controlled by nonuniform and highly heterogeneous stresses, associated with the deformation style of Lake Van. The frequent redistribution of flickering stresses and nonlinear deformations in the rupture area increase the *b*-value and decrease the radiation efficiency.

Keywords: eastern Anatolia, Van earthquake, aftershocks, seismic *b*-value, radiation efficiency, micromechanisms, nonlinear deformations

1. Introduction

Eastern Anatolia is a seismically active accretionary complex region where the Arabian and Anatolian plates collide (**Figure 1a**). The Anatolian-Arabian plate collision takes place along a deformation zone, the Bitlis Thrust Zone (BTZ) (**Figure 1a**) [1–3]. A westward extrusion of the Anatolian plate is along the two major strike-slip faults: the dextral North Anatolian Fault (NAF) and the sinistral East Anatolian Fault (EAF) zones. These faults join each other at the Karlıova Triple Junction (KTJ) in eastern Anatolia (**Figure 1a**). The east-west trending Mus-Lake Van (MRB) ramp-shaped lake basin is a conspicuous tectonic feature in the region and roughly reflects the N-S compression and W-E extension (**Figure 1a**) [1].

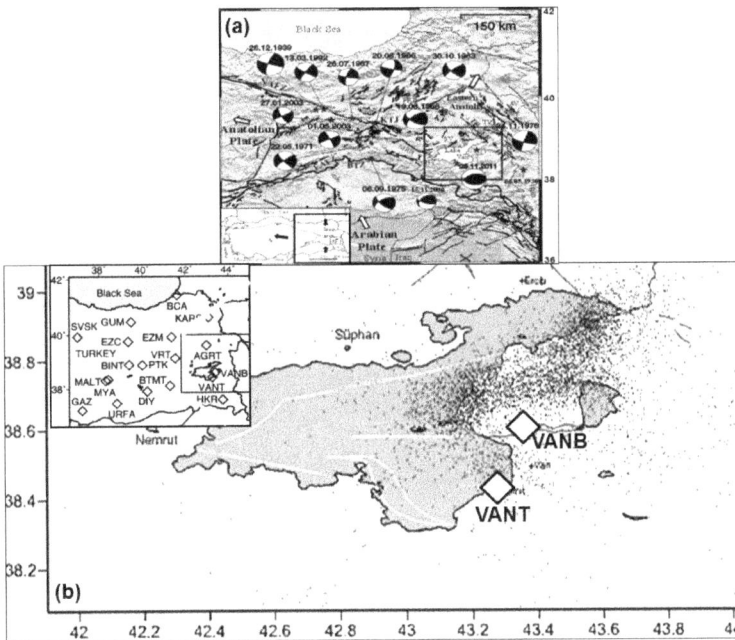

Figure 1. Major tectonic elements and seismicity of eastern Anatolia. (a) Major tectonic elements of eastern Anatolia with the *Ms* ≥ 6.0 earthquakes epicenters (stars) and known focal mechanisms (compiled from [4–8]. KF, Karayazı Fault; TF, Tutak Fault; BTZ, Bitlis Thrust Zone: MRB, Mus Ramp Basin; NAFZ, North Anatolian Fault Zone; PRB, Pasinler Ramp Basin; KTJ, Karlıova Triple Junction; EAFZ, East Anatolian Fault Zone; EF, Ercis Fault; HT, Hasan Timur Fault; AV, Ağrı Volcano; NV, Nemrut Volcano; TV, Tendürek Volcano; SV, Süphan Volcano. A large rectangle with large arrows encloses the map shown in **Figures 2** and **3** indicate relative plate motions. Map reprinted from Ref. [9] with kind permission of Springer Science and Business Media. (b) Epicenters determined by the KOERI network (inset map) are the basis of our analysis. The inset map shows seismographic stations used for reexamination of hypocenters in this study. A rectangle indicates the study area of this paper.

Two intraplate (midplate-related) earthquakes took place within an accretionary complex area, near the Lake Van basin (**Figures 1** and **2a**), the former 23 October 2011 Van event (M_w, 7.1) occurred to the east of Lake Van and the later 9 November 2011 Edremit event (M_w, 5.6) occurred to the south of Lake Van (**Figure 2a**). The preliminary source mechanism solutions of these earthquakes indicate almost pure thrust and oblique-slip faulting, respectively (**Figures 1** and **2a**). The epicenters of the two earthquakes were reported by Kandilli Observatory and Earthquake Research Institute (KOERI, Turkey) and by USGS (**Figure 1b**). The hypocentral and source parameters of these earthquakes estimated by different organizations and seismological institutes are summarized in the map views given in **Figures 2** and **3**.

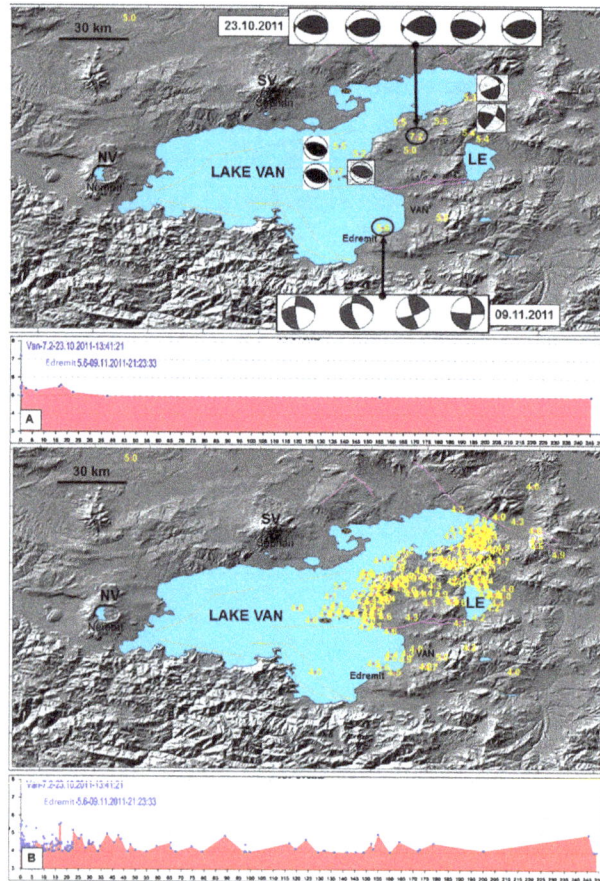

Figure 2. Aftershock seismicity and faulting of Lake Van basin. Topographic maps of the Lake Van area and vicinity showing the epicenter distribution of aftershock events (5487 events) between 23 October 2011 and 13 April 2012 (173 days) (LE: Lake Erçek). Boundary faults of Lake Van basin are also shown in the maps (see the text for details). The main 14 aftershocks with magnitudes $M_w \geq 5.0$ as a function of time (245 days) are shown in A. The 189 aftershock events with magnitudes $M_w \geq 4.0$ as a function of time (248 days) are shown in B. The locations of the 2011 Van (M_w, 7.1) and Edremit (M_w, 5.6) earthquakes and their focal mechanisms are indicated in a, by different institutions (KAN, USGS, EMSC, AZUR, GFZ, ERD, HARV, INGV). (KAN: B.U. Kandilli Observatory and Earthquake Research Institute (KOERI); EMSC: European Middle East Seismology Center; AZUR: Nice University, GeoAzur Laboratory, France; GFZ: Geoforshung Zentrum, Potsdam, Germany; ERD: Disaster Management and Emergency Presidency, Ankara, Turkey; HARV: Harvard CMT; INGV: Insituto Nazionale di Geofisicae Vulcanologia, Italy; USGS: United States Geological Survey.

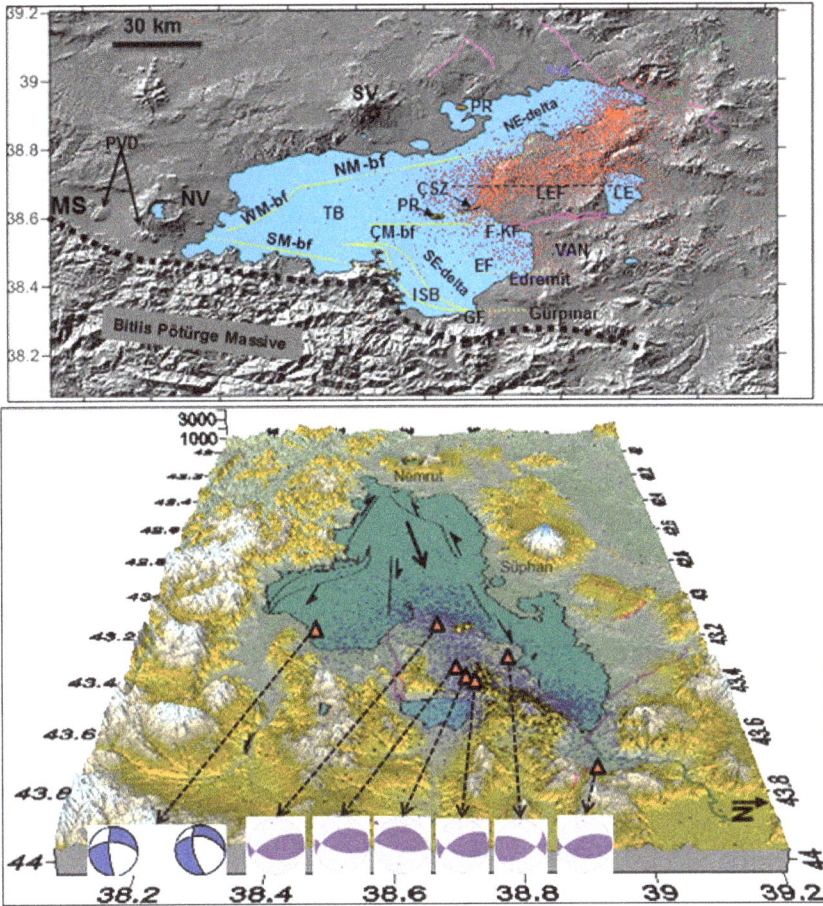

Figure 3. Aftershock seismicity, focal mechanisms, and faulting of Lake Van basin. Yellow lines demarcate the margin boundary faults and red dots indicate the aftershock distribution in the major basinal provinces of Lake Van basin. NM-bf, Northern Margin Boundary Fault; WM-bf, Western Margin Boundary Fault; SM-bf, Southern Margin Boundary Fault; ÇM-bf, Çarpanak Margin Boundary Fault; TB, Tatvan Basin; PR, Pressure Ridge; ÇSZ, Çarpanak Spur Zone; ISB, Internal Sub-Basin; F-KF, Fault-Kalecik Fault; LEF, Lake Erçek Fault; LE, Lake Erçek; GF, Gevas Fault; SV, Süphan Volcano; NV, Nemrut Volcano; PVD, Parasitic Volcanic Domes; MS, Mus Suture (black-dashed line) [12, 13]. 3D digital elevation block diagram of Lake Van basin shows relative fault motions and locations of some distinct events with fault focal mechanisms, as shown in **Figure 2**.

The Van earthquake (M_w, 7.1) was the largest thrust earthquake known to have occurred in the Van area (**Figure 2a**) and Turkey since at least the 1976 Çaldıran-Muradiye event of Ms, 7.3 [9–11] (**Figure 1a**). The Van earthquake was related to the oblique "blind" thrust faulting, which is inconsistent with mapped boundary faults in the Lake Van area [12–14]. Multifractal occurrence and distribution of long-period aftershock activity and focal mechanism of this larger event show a NE-SW striking rupture plane dipping toward northwest [15–17]. The rupture gradually expanded near the hypocenter and propagated both northeast and southwest, but mainly to southwest [15]. The 9 November 2011 (M_w, 5.6) Edremit earthquake (5–7 km depths) occurred offshore to the south of Van along the north dipping a normal oblique-strike-slip Edremit fault [9] (**Figure 2a**). The epicenter locations and the fault focal solutions of these earthquakes indicate that they occurred on different faults (**Figures 1b** and **2a**).

Seismic scaling relation shows that the way of quantifying the mechanical efficiency of earthquakes is with the radiation efficiency parameter (η_R) as defined in Ref. [18]. The smaller radiation efficiency ($\eta_R < 1.0$) or higher critical fracture energy [19] has an important implication for the rupture growth of the seismic activity and is easily comparable with the high b-values ($b > 1.0$) and time-dependent variability in the b-value. For the 2011 Van earthquake to determine the seismicity changes and understand the local stress state, the irregular changes in b-value were reported to be extremely localized spatially [20]. This case requires subsurface stress perturbations occurring at similar and smaller spatial scales. The 2011 Van earthquake combined with intense fluid withdrawal and gas migration [14, 21] may cause frequent redistribution of flickering stresses [22]. This leads to nonlinear effects such as the rapid development of an additional fault-fracture complex in Lake Van. This, in turn, facilitates the subsidence process [23, 24], increases the b-value ($b > 1.0$), and decreases the radiation efficiency ($\eta_R < 1.0$) in the mainshock area [24]. This inverse relationship means that the higher b-value causes the lower radiation efficiency and vice versa. Since, the massive fracturing and flickering stresses induce small-scale faulting movements and nonlinear deformations, which produce smaller earthquakes (microseismic activity) in the area of the mainshock and its vicinity. These nonlinear deformations contain elastoplastic and viscousplastic (or hydroelastic) components, which can change with time and depth, and the rate and the total amount of seismic deformation energy [22–24] and thus, control both the b-value and the radiation efficiency (η_R).

The aftershocks shown in **Figures 1b** and **2** associated with the 2011 Van mainshock still continue to occur today and there were several thousands of events from October 2011 to July 2012. In this study, it can be found that the aftershocks of the Van mainshock play a key role in the appearance of strong seismic coupling between basin-bounding faults, faulting style, and active deformation of Lake Van (**Figures 2** and **3**) and upper crustal seismogeneity [12–14, 25, and the references therein]. This allows for a detailed investigation of thousands of events in the focal region and provides an objective path to obtain an improved understanding of the stress state of the 2011 Van earthquake. Aftershock magnitudes of $M \geq 4.0$ observed in the 2011 Van event (**Figure 2**) may suggest that an amount of energy is not radiated and used in the form of fracture and frictional energy near the focal region [19]. Some of the nonradiated energy may also function to significantly increase the temperature in the focal region [16, 19, 26, 27]. The b-values calculated for various sizes of events in the 2011 Van earthquake suggest both high and low b-values. This variability in b-values is considered to support the idea that some part of the volcanic intraplate crust may have increasing mechanical heterogeneity (**Figure 4**). This may decrease the shear stress after the Van event. To test these concepts, imaging time and depth-dependent variability in b-value and correlating aftershock seismicity with b-values seemed to be an appropriate approach. In this study, a first attempt is made to interpret the variations in the b-value combined with the radiation efficiency (η_R) and the cross-sectional profiles of hypocenters of aftershocks. Thus, the current research of the 2011 Van earthquake aims to investigate the time-dependent distribution of seismic b-values of aftershocks in the light of the active tectonic deformation of Lake Van and provide a new view of the seismic deformation characteristics of this larger dip-slip event.

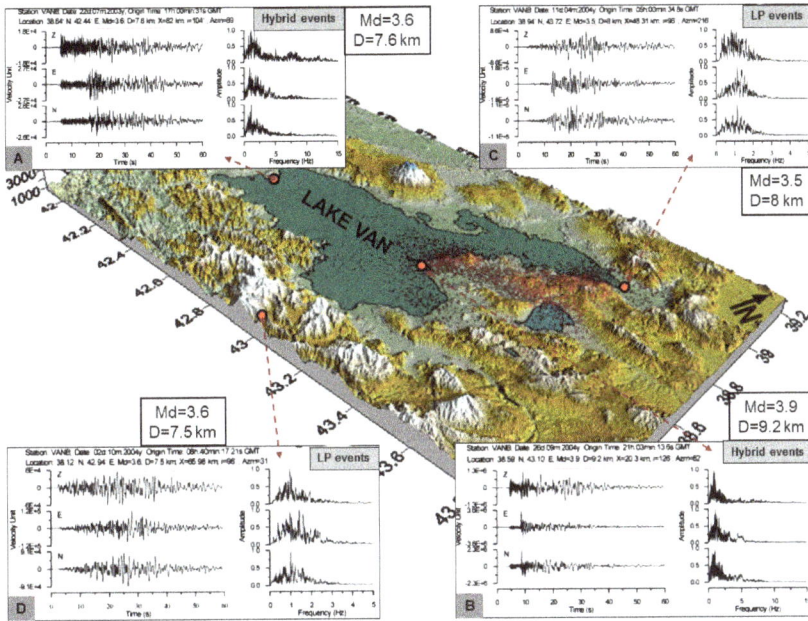

Figure 4. Velocity seismograms of hybrid and long-period (LP) events at the VANB broadband station. Examples of the three-component velocity seismograms of hybrid (A and B), LP events (C and D) recorded at the VANB broadband station and their normalized amplitude velocity spectra. Dashed arrows show the locations of classified earthquakes. (Md: duration magnitude, D: depth (km), X: distance from hypocenter to receiver (km), i: incidence angle measured from downward vertical (°), Azm: azimuth angle (°). Earthquake waveforms reprinted from Ref. [28] with kind permission of Springer Science and Business Media.

2. Structural setting and tectonic elements of Lake Van Basin

The Lake Van basin is the eastern continuation of the MRB and was separated from it by the Nemrut Volcano (NV) with parasitic volcanic domes (PVD) [1, 2] (**Figures 1a** and **3**). Deformational patterns of the lake have been formed as a result of the tectonic structure of eastern Anatolia [14, 29]. Lake Van was formed through a combination of normal and strike-slip faulting and thrusting [1, 2, 29]. The strike-/oblique-slip deformation in Lake Van caused distinct strike-slip sedimentation, extensional magma propagation through boundary faults (see **Figure 2**), upper crustal seismicity, and hydrothermal activity [12–14, 29–33].

The multichannel seismic reflection profiles collected from Lake Van basin [12–14, 21, 25, 34, 35] and bathymetry data [36–38] revealed that the lake basin is dominated by a deep Tatvan basin (TB), north-eastern delta (NE-delta), and a south-eastern delta (SE-delta) (**Figure 3**). The lake is completely bounded by steep, oblique boundary faults, namely the northern boundary fault (NM-bf, transpressive), southern margin boundary fault (SM-bf, transtensive), western margin boundary fault (WM-bf, transtensive), and Çarpanak margin boundary fault (ÇM-bf, transpressive) (see **Figure 3** for the sense of shear). In the current study, ÇM-bf is currently named because of its prominent location close to Fault-Kalecik Fault (F-KF) or Van Fault [39]. These marginal faults have a basinward, downthrown side (TB in **Figure 3**), and gently folded sedimentary sections on the downthrown side of these faults together with some small splay faults (**Figure 3**).

The location of the 2011 Van mainshock and epicentral distribution of its aftershocks (**Figures 1** and **2**) occurred in a wedged-shape area (**Figure 3**) that is transpressively uplifted by major boundary faults, namely ÇM-bf (F-KF) in south and NM-bf in north (**Figures 2** and **3**). These faults are prominent structural elements, controlling the dextral faulting regime in the east of the lake and also the 2011 Van mainshock. Ref. [40] gave field evidence for possible dextral faulting, extending roughly in the E-W direction, east of the eastern shore of the Lake Van basin (**Figure 3**). These faults have been mapped by several studies such as Lake Erçek Fault (LEF), Kalecik fault [40, 41], inferred Edremit fault (EF) [40, 42], Gevas fault (GF) [43, 44] (**Figure 3**). These faults are dextral oblique-slip faults lying east and south-east of the Lake Van basin [9, 45, 46].

High heat flows, hydrothermal discharges, and CO_2-emissions in Lake Van may give some evidence of seismic ductile events (high b-value) in the upper crust beneath the lake. Ref. [28] observed three types of seismic events (≤ 10 km) on the spectral band of earthquake waveforms in and around the Lake Van area: hybrid events, long-period (LP) events (see **Figure 4**), and a kind of tremor. In the study, Ref. [28] revealed the presence of upward rising of magmas and the instability of high viscous lava domes [47] in and around the Lake Van basin. The calculated high b-values in this study were associated with an increase in material heterogeneity [48] and thermal gradient [28, 49] near the Lake Van area.

3. Data and methodology

3.1. Used data

The earthquake catalog published by Kandilli Observatory and Earthquake Research Institute [50] of Turkey was used to determine the seismicity pattern of aftershock events as a function of time and develop the clean images of the seismic b-values. Fifteen permanent broadband seismic stations with high-gain seismometers provided real-time data (**Figure 1b**). After the Van event, eight temporal seismic stations were operated by KOERI around the Lake Van area. The location errors in the focal depths of the aftershocks are about 2–3 km [17, 20, 27, 51]. We did not relocate the focal depths of aftershocks and only precisely located focal depths of aftershocks were used in our study (see Refs. [17, 20] for the criteria used to select high-quality events and the focal depths). The uncertainties in the focal depths are considered not to affect the b-value estimates and the main results in this study. The earthquake catalog of the study area includes Lake Van basin, spanning the period 23 November 2011 to 5 July 2012 and contains 6005 events over 255 days. The depth and magnitude of the earthquakes range from 5 to 30 km and 1.5–7.1 M_w, respectively. To establish cross-sectional images of the hypocentral depths and epicenter distribution of the aftershocks, we selected 5304 event epicenter distributions for a time period from 23 November 2011 to 28 March 2012 (**Figures 1b** and **3**).

3.2. Frequency-magnitude relationship

The frequency-magnitude distribution (Eq. (1)) defines the relationship between the frequency of occurrence (foo) and the magnitude of earthquakes. The size distribution of earthquakes is adequately described by the G-R relation [52]:

$$\log_{10}N\ (\geq M) = a - bM \tag{1}$$

where $N(M)$ is the frequency of earthquakes with magnitudes higher than or equal to M, and the parameter a is a real constant that characterizes the seismic activity. The b-value varies regionally, both spatially and with depth and describes the relative size distribution of events.

The frequency-magnitude distribution (Eq. (1)) and the variation in the b-value depend on the tectonic and rheological conditions such as material heterogeneity, crack density, thermal gradient, creep, applied stress, and asperities [48, 53, 54]. The analyses of laboratory data and observed seismicity suggest that the b-values are related to strength heterogeneities and the state of stress on a fault [48, 53, 55, 56]. The estimated b-values of regional and global seismicity also exhibit systematic dependencies on depth, focal mechanisms, and other statistical variables [16, 27, 57–61].

Although the b-values equal 1 over long-timescales and large spatial scales, significant variations occur on smaller scales. A b-value greater than 1.0 is related to the areas of crustal heterogeneity and low applied stress [61–63]. However, the b-value lower than 1.0 indicates high differential stress [61–63] and volumes with crustal homogeneity [64–66].

In this study, the b-value was estimated by the maximum-likelihood (ML) method [67] and calculated using ZMAP software [68]:

$$\log_{10}e = b\ (M_{mean} - M_C) \tag{2}$$

where M_C is the minimum (cutoff) magnitude of the given sample [69], M_{mean} is the mean magnitude, and the variable "e" is the Napier's number (base of the natural logarithm), which is approximately 2.7183. This formulation is described by a linear relationship with the constants a and b. It is generally about 1 depending on the number of aftershocks in the time and region sample (a) and the slope (b). The shorter the time and/or the smaller the area, the more the fit is degraded by the statistics of small numbers [70]. In a statistical study conducted by Ref. [27], about 5454 aftershock events were obtained by sampling longer intervals (256 days) in a large epicentral area (the Lake Van) and this produced better fits (see Ref. [27]). Although the aftershock data in that study were generally well described by the linear relationship, some deviations were observed. The data deviated from the $b = 1$ line for very small ($M_C < 2.5$) magnitudes and the aftershock events with $M_C > 2.5$ closely followed the G-R relation with $b = 1$. This result suggests that the G-R relation and Omori-Utsu law provide good overall descriptions and estimations of aftershock seismicity [27].

In this study, the standard deviation (δb) in the b-value was estimated using Eq. (3) devised by Ref. [69] and modified by Ref. [71]:

$$\delta b = 2.3 b^2 \sqrt{\frac{\Sigma \left(M_i - \{M\} \right)^2}{n(n-1)}} \tag{3}$$

where n is the sample size.

Detectable minimum magnitude (M_C) is a necessary parameter using with the maximum number of aftershock events over a long period of time to achieve a more accurate analysis [72, 73]. The variation in M_C as a function of time during the period 23 October 2011 to 1 August 2012 was shown in Ref. [27]. M_C irregularly fluctuated due to the large number of aftershock events, and the strong changes in seismicity.

The depth-dependent variations of the b-value were imaged using the focal depths of events (5 days × 5 km), which were densely clustered from 5 to 15 km. The depth cutoff technique is not used; however, focal depths more than 15–25 km are not taken into account in this study. We chose to use a time interval approach to image the b-value. The variation in the b-value as a function of time and focal depth is imaged for sampled time interval (170 days) including 5454 aftershock events (**Figure 5**). To perform high-resolution images of b-values, we tried different time intervals of 70 days and increased the window to 1930 events for the interval between the 20th and 90th days (**Figure 6a**) and to 855 events for the interval between the 100th and 170th days (**Figure 6b**). This was required due to the distinct variation in the foo of the observed events as a function of time (cumulative number of events/per days).

3.3. Seismic scaling relations

Static stress drop $\Delta \sigma_s$ is roughly constant over a range of M_0 from 10^{18} to 10^{23} N m [19]. Ref. [17] estimated that in the 2011 Van earthquake with its 18 km focal depth, the $\Delta \sigma_s$ is, on average, approximately 10–15 MPa. This range is roughly consistent with the assumption that the rupture-front velocity (V_r, 3.2 km/s) is, on average, similar to the S-wave velocity (V_s, 3.9 km/s) of the Van earthquake (see Ref. [17] for the velocity model). The scaled energy, $\tilde{e}$, is a dynamic parameter that can be determined by the ratio of the radiated seismic energy to seismic moment;

$$\tilde{e} = E_R / M_O \tag{4}$$

This ratio can be interpreted as proportional to the energy radiated per unit fault area and per unit slip [74, 75]. In this study, following the approach of Ref. [19], $\tilde{e}$ is taken as a constant, because the static stress drop, $\Delta \sigma_s$, is constant. E_R is difficult to determine accurately due to the complex wave propagation and scattering effects of the 2011 Van earthquake. It is estimated that $\tilde{e}$ is, on average, approximately 5×10^{-5} with a range 10^{-4}–10^{-6} for large earthquakes ($M_w \approx$ 7.0) [19, 76, 77].

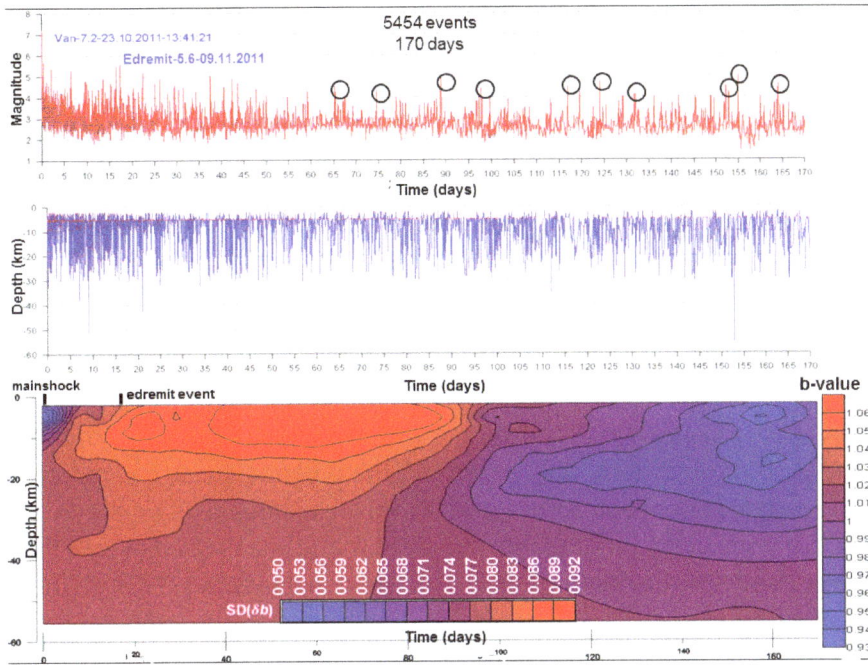

Figure 5. Time and depth-dependent variation in the b-value and aftershock seismicity pattern. Time and depth-dependent variation in the b-value is correlated with aftershock seismicity that indicates the time-dependent variation in magnitude and focal depths of 5454 events for 170 days. Standard deviation (SD) in the b-value falls in the range of 0.05–0.09. Dots show the distinct events with magnitudes ($M_w \geq 4.0$). Note the time-dependent relation and compatibility between variation in the b-value and magnitudes of events. This indicates that a relative increase in the number of the events ($M_w \geq 4.0$) is compatible with the decrease in the b-value (see the first days).

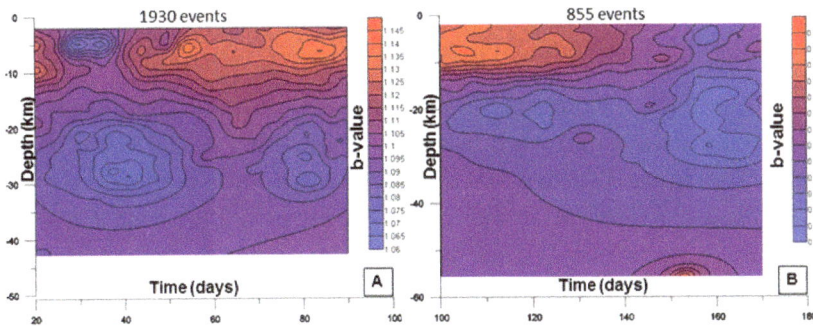

Figure 6. High-resolution images of time and depth-dependent variation in the b-value. High resolution images of time and depth-dependent variation in the b-value given in A for the calculated 1930 events between the 20th and 90th days ($b > 1$) and in B for the calculated 855 events between the 100th and 170th days ($b < 1$).

The radiation efficiency, η_R, determines the dynamic character of an earthquake [18, 19, 77]. A relation between η_R and observable seismological parameters was obtained, thus:

$$\eta_R = 2\mu \tilde{e} / \Delta\sigma_s \qquad\qquad (5)$$

where, $\tilde{e}$ is 5×10^{-5} ($\tilde{e}$ is always less than 10^{-4}) [19, 78], $\Delta\sigma_s$ represents the static stress drops with a range 10–15 MPa, and μ is a shear modulus, 3×10^4 MPa for the 2011 Van earthquake. The seismic moment M_0 5.37×10^{19} N m (M_w, 7.1 and the centroid focal depth 18 ± 2 km) in the waveform fit using a shear modulus μ of 30 GPa (the rigidity of the top 15 km) is given by the velocity model proposed by Ref. [17] for the Van earthquake (see [17]). Using the estimates of these parameters, the radiation efficiency, η_R, is determined for the 2011 Van earthquake using Eq. (5).

The computed η_R values of the 2011 Van earthquake typically fall in the range of 0.2–0.3 within a range of 10–15 MPa ($\Delta\sigma_s$). It is also estimated that the η_R value is 0.6 for a value of 5 MPa ($\Delta\sigma_s$) (data from Ref. [17]). The radiation efficiency of most earthquakes lies between 0.25 and 1. Ref. [19] classified the radiation efficiency for the rupture zone interpretation as follows:

If $\eta_R = 1$, the breakdown zone is unimportant and failure occurs primarily in the steady-state regime. No energy is mechanically dissipated and the potential energy is radiated as seismic waves, after the heat loss has been subtracted, and the earthquake is considered to be a very brittle event. If $\eta_R = 0$, no energy is radiated, the event is quasi-static and even if the static stress drop is very large as in the range of 10–15 MPa estimated in this study. If $\eta_R \ll 1$, the microscopic breakdown process dominates the dynamics.

4. Results

4.1. The variation in the b-value as a function of time

In this study, the estimated b-value varies from 0.93 to 1.07 for 5454 events (**Figure 5**), 1.06 to 1.15 for 1930 events, and 0.85 to 0.99 for 855 events (**Figure 6**). The estimated b-values generally fall in the range 0.85–1.15 with standard deviations (δb) in the range of 0.05–0.09 (**Figure 5**). The range of the standard deviation (δb) is relatively low.

The resulted variation in the b-value clearly exhibits two dramatic changes in the b-value, one ($b > 1$) during the first 100 days of the mainshock and the other ($b < 1$) in the last 70 days of the mainshock (**Figure 5**). **Figure 5** shows that the event magnitude begins to drop considerably, corresponding to the time interval of the high b-value, and after that it continues to drop, corresponding to the time interval of the low b-value and it decreases significantly. These remarkable changes can be seen to coincide with the 23 October 2011 (M_w, 7.1), Van and 9 November 2011 (M_w, 5.6), and Edremit events (**Figure 2a**).

The variation in the b-value after the mainshock ranges from 0.93 to 1.06 for 5454 events over 170 days, suggesting a relatively small change in b-value (**Figure 5**). After a large number of aftershocks, the b-value began to drop slightly, reaching a normalized constant value. The constant b is typically equal to 1.0, indicating a seismically active time interval (**Figure 5**). This interval may indicate a transitional variation in the b-value. Then, the b-value continued to drop, reaching its lowest value ($b < 1$ in **Figure 5**). **Figure 5** shows the b-value ranging from 1.06 to 1.0 over the first 100 days and from 1.0 to 0.93 over the last 70 days. High b-values suggest that a large number of small or same-sized aftershocks of the Van and Edremit events are dominant. However, low b-values suggest a relative decrease in the smaller magnitudes of events and an increase in the medium magnitudes of events (**Figure 5**). This reveals that the time-dependent variation of individual events with magnitudes ($M_w \geq 4.0$) dominantly resulted in low b-values.

Time and depth-dependent differences in b-values illustrate significant seismic variability (**Figure 6**) for different intervals. We clearly observed a sharp decrease in the foo, from 300 to 100 events, between the 20th and 25th days, typically giving a "sharp stress drop pattern" of intraplate seismicity [27, 79, 80]. We selected the 20th day as a starting point, extending to the 90th day (**Figure 6a**). After the sharp pattern of stress drop, the selected interval between the 20th and 90th days (**Figure 6a**) seems to be well suited to reveal and understand the superimposed effects of the first (Van) and the second (Edremit) mainshocks on the variation in the b-value. From the 90th day, the relative number of individual events with magnitudes of $M_w \geq$ 4.0 considerably increases toward 170th day (**Figure 6b**). The Van mainshock was followed by a b-value increase near the source area during approximately 100 days after the mainshock (**Figure 6a**). The variability in the b-value for time interval between the 20th and 90th days was between 1.06 and 1.14 (**Figure 6a**) for 1930 events, which suggests $b > 1$, while the variability ranges from 0.85 to 0.98 for 855 events suggesting $b < 1$, and for the time interval between the 100th and 170th days it was between 0.85 and 0.98 (**Figure 6b**). **Figure 6a** shows the increase in the b-value for the period between the 20th and 90th days and it is followed by a slight decrease in the b-value, to 0.85 minimum (**Figure 6b**). The zone of the low b-value (0.85–0.98) anomaly (**Figure 6b**) was found in the vicinity of the epicenter of the mainshock (**Figure 2a**). This low b-value in the focal area indicates that medium-sized events are likely to take place (**Figure 2b**).

As shown schematically in **Figure 7**, the high and low b-values illustrated for the 5454 events over 170 days (**Figure 5**) correlated with high-resolution images of variation in the b-value for 1930 events and 855 events (**Figure 6a and b**, respectively). The schematic illustration of the variation in the b-value shows that the b-values are consistent with basin-bounding faults, fault focal mechanisms, and the magnitudes of the events (**Figure 3**). In **Figure 7**, the observable increase in the b-value is correlated with lowering the effective stress level. This indicates increased material heterogeneity and reduced shear stress [81]; however, the dramatic decrease in the b-value is correlated with rising effective stress level prior to major medium-sized earthquakes [82], indicating high stress accumulation and an increase in shear stress [81].

4.2. The frequency-magnitude statistics of the aftershocks

We estimated the b-values for various sizes of aftershocks and then showed the obvious anomaly of aftershock seismicity pattern. Our results mainly relate the variations in the b-value to the stress state and temporal stress transfer in and around the focal area.

The variation in the magnitude of completeness (M_C) (23 October 2011 to 1 August 2012) irregularly fluctuated due to the large number of aftershocks, and the rapid changes in seismicity (see data in Ref. [27]). A moving window approach was chosen to create a temporary change in M_C, with a window size of 5454 events to observe the M_C for the sequence (256 days), using the maximum curvature [27]. The variations in the minimum magnitude (M_C) for the entire catalog ranged from 1.4 to 2.3. The M_C for the aftershock sequence was estimated to be equal to 2.5 with a 95% goodness-of-fit level [27].

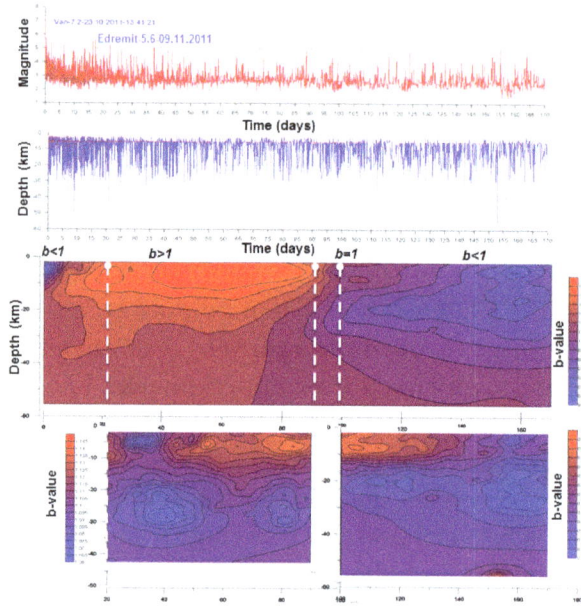

Figure 7. Correlation of the variation in the *b*-value with aftershock seismicity. Correlation of the time and depth-dependent variation in the *b*-value with aftershock seismicity given in **Figure 5** and the high-resolution images of variation in the *b*-value given in **Figure 6**. White-dashed lines show time intervals for *b*-values.

Ref. [27] calculated the exponential and linear plots of the *b*-value for events larger than M_C with one population; the G-R distribution of events smaller than 7.1. The G-R statistics ranged over the magnitude interval $M_C < M < M_{mean}$. For the 5454 aftershock events, there was a variation in the *b*-value ranged between magnitudes 2.5 (M_C) and 7.1 (M_{mean}). The *b*-value for this sequence was estimated to be equal to 1.9 ± 0.05 with a 95% goodness–of-fit level [27]. This supports the idea that the aftershock events follow the G-R frequency-magnitude statistics of events [52].

Ref. [20] reported that M_C for the Van sequence was estimated to be equal to 2.6 with 90% goodness-of-fit level and that the *b*-value for this sequence was estimated to be equal to 0.89 ± 0.02 with a 90% goodness-of-fit level (see Ref. [20] for the spatial variations of M_C). Because the *b*-value is lower than the global mean value of 1.0, Ref. [20] reported that the Van sequence consisted of larger magnitude aftershocks and high differential crustal stress in the regime. However, this result is considerably different from the result from the current study due to the higher *b*-value ($b > 1.0$). The higher *b*-value indicates that the Van aftershock sequence consists of smaller and/or medium-sized magnitude aftershocks and relatively low differential crustal stress in the focal area [61, 83].

In the previous studies, the observed b-values of regional seismicity fall in the range 0.7–1.3 [57–59] and in the range 0.7–1.2 [20] for the Van sequence. The cross-sectional maps of depth distribution of the b-value performed by Ref. [20] fall in the range 0.8–1.65. The b-values vary from 0.9 to 1.5 along the Van rupture fault zone computed by Ref. [84]. These results clearly indicate that the estimated b-values with the low (δb) in this study are statistically significant. The estimated b-values in this study are also correlated with the slip distribution models of the 2011 Van mainshock obtained through the inversion of seismological and geodetic data from Ref. [84]. Thus, we can explain a possible physical process over the source fault or surrounding crustal volume that may control the b-value as a function of time.

4.3. The hypocentral cross sections

A prominent aseismic zone can be observed in the deep central basin of Lake Van (TB), where no remarkable events are recorded (**Figure 3**). Aseismicity suggests that there is clear evidence of seismic quiescence (SQ) in this area (TB) (**Figure 3**). This may imply a strong structural asymmetry between the seismic (ÇSZ-LEF-LE) and aseismic zones (TB) in Lake Van (**Figure 3**). This asymmetry is also illustrated by the depth distribution of the epicenters at upper crustal depths (**Figure 8**). In the hypocentral cross sections in **Figure 8**, aseismic (TB) and seismic zones (ÇSZ-LEF-LE) with strong crustal asymmetry are referred to as seismic quiescence and uplifting (U) zones, respectively (which can be seen by comparing **Figure 3** with **Figure 8**). The aseismic (TB) or seismic quiescence zones are distressed and declustered, while seismic zone (ÇSZ-LEF-LE) or the uplifted (U) area are highly stressed and clustered. The SQ zone extends over a thick sedimentary volume of TB surrounding the focal zone, while the U zone is rooted in a fault-bounded block basement (ÇSZ). Previous seismic reflection studies reported that the deep central basin (TB) of Lake Van is thick depositional province (600 m), onlapping onto the shallowing delta settings (NE-delta), toward ÇSZ where the deposition is sharply terminated, reaching a thickness of sediments of 150–250 m. Sediment termination is occurred due to the fault-bounded block uplift (ÇSZ) [12–14]. This finding reflects a typical pattern of strike-/oblique-slip deformation and sedimentation [1, 2, 85] and is in good agreement with the high and low b-values (**Figure 7**) and the seismic asymmetry (**Figure 8**).

Time-dependent variability in the b-value from high to low (**Figure 7**) and the aseismic TB (**Figure 3**) may suggest a low b-value with the occurrence of medium-sized events near to the focal area in future (events with $M \geq 4.0$ in **Figure 5**). This assumes a basinal stress transfer [79] from TB through the boundary faults toward ÇSZ and the focal area (**Figure 3**). The low b-value is interpreted as there being a complex fault network area, highly strained (or locked) and a stress condensation zone nearby the focal area (**Figures 2** and **3**). This zone may indicate asperities on the complex faults prior to a large dip-slip earthquake [79]. This complex zone seems to be an isolated fragmentation barrier [86], implying a strong seismic coupling between the Lake Van basin and the focal area (**Figures 2** and **3**). This barrier is the probable zone for future medium-sized earthquakes (**Figures 2a** and **5**). This suggests that high stresses are being accumulated and condensed, causing a potentially locked zone near the focal area.

These results indicate that the variation in the b-value for 170 days implies a highly fractured and heterogeneous earthquake setting. The 2011 Van earthquake and its aftershock distribution were controlled by irregular stress and strain patterns in and around the focal zone, associated with the present-day deformation style of Lake Van basin (**Figure 3**). This suggests that remarkable physical and mechanical changes occurred in the seismic deformation potential of the crust beneath the Lake Van basin.

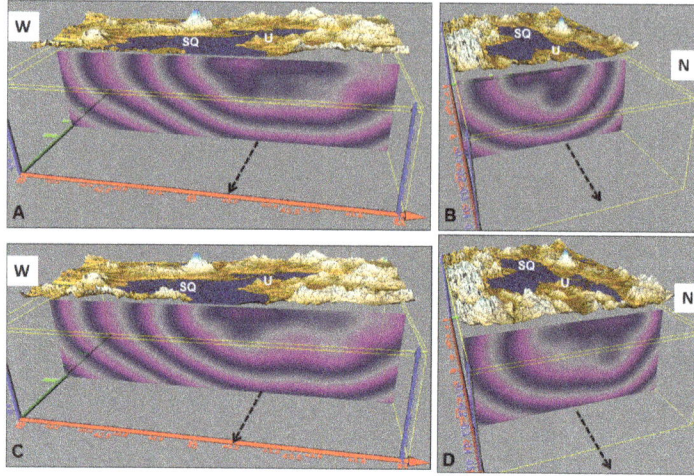

Figure 8. Hypocentral profiles of aftershock events. Low-resolution cross-sectional images of the hypocenter distribution of 5304 events indicate radiated patterns of seismic density of events at crustal depths shown by sections projected on the shaded relief map. W-E projections are shown by A and C and N-S projections are shown by B and D. W-E and N-S cross-sections seen in C and D are the sections toward the south. Directions are indicated by black-dashed arrows to show seismic density changes in different projections. Note that the deep central basin of Lake Van is aseismic, indicating seismic quiescence (SQ) zone and the focal area is seismic with high event density, indicating the uplifted (U) zone (see **Figure 3**) (depths in km × 100).

5. Interpretation and discussion

The time-dependent variability of the b-values (**Figure 7**) of the 2011 Van earthquake provided useful and significant information about the seismic deformation characteristics of the Van mainshock. The computed b-value distribution as a function of time over the Van focal area underwent significant variation throughout the aftershock sequence. The Van event was followed by aftershocks of variable medium magnitudes. The data given in **Figures 2** and **3** show that aftershock seismicity is concentrated in time and space in relation to the principal mainshock event, growing nonlinearly and widely varying in space and time.

The resulting variation in b-value can be seen to coincide with the two mainshocks of the Van (23 October 2011) and Edremit (9 November 2011) events (**Figure 2a**). In this study, the variability in the b-value is correlated with decreasing and increasing effective stress levels. The basinal distribution of events, hypocentral cross sections, active tectonics, and deformation style of the Lake Van basin suggest that differing amounts of rapid stress changes over 170 days or the type of driven mechanism at crustal depths caused the variations in the b-value (**Figure 7**).

5.1. Stress state as a function of time

The magnitude characteristics of the aftershocks, including the low b-values (**Figure 7**), are strong evidence of the cause resulting from high stress accumulation. The individual events with magnitudes of $M_w \geq 4.0$ as shown in **Figure 5** can be interpreted as indicating the potential zones of locked faults and reducing the b-value. However, the magnitude characteristics of the aftershocks, including the high b-values (**Figure 7**), are being caused by extreme crustal heterogeneity. The large number of smaller events with magnitudes of $2.0 \leq M_w \leq 4.0$ can be interpreted as indicating a decrease in shear stress and increasing the b-value. The strong clustering of aftershocks with time and variable b-values over 170 days fits with an inhomogeneously stressed crust with high fracture energy. The high amount of fractured energy reduces the strength, causes low shear stress (high b-value), and allows the stress to be released [19], but the fracture energy can also create locked seismic zones (low b-value) (see events $M_w \geq 4.0$ in **Figure 5**).

High-resolution images of the b-values for time intervals from the 20th to the 90th day (**Figure 6a**) and the 100th to the 170th day (**Figure 6b**) are shown in **Figure 7**. The high b-value over 70 days may have some similarities to the swarms [87]. However, the low b-value indicates locked fault zones. As shown in **Figure 7**, the Van and Edremit events considerably lowered the general stress in the area (see Ref. [79] for the amount of stress drop). The smaller magnitudes are dominant until the stress builds up again (**Figures 5** and **6a**). Then, the medium magnitudes are dominant, reducing the b-value (**Figures 5** and **6b**). In **Figure 7**, the difference in b-values from the 20th day to the 90th day and the 100th day to 170th day period may be an indicator of a situation where there is stress transfer from a lower to a higher stress. This difference can be interpreted as the stress redistribution of the Van aftershock sequence [27].This stress pattern indicates that the aftershock events (the October 2011) may have promoted and triggered the events (the November–December 2011 and the January 2012). This result is consistent with the Coulomb stress variation in the entire region of Lake Van [20, 88] and temporal and spatial distribution of the b-values [84].

5.2. The variability in the b-value and the radiation efficiency (η_R)

As stated above, the high b-values in **Figure 7** indicate a general decrease in shear stress and increasing complexity in and around the focal zone. There is a slight variability in the b-value from high to low (**Figure 7**). This suggests an increase in the stress level, from low shear to high effective stress. The stress is transferred from the interval dominated by high material heterogeneity and fracture energy to the interval dominated by high stress accumulation and strain condensation.

Previous studies have stated that a large number of smaller and/or same-sized aftershock events with a long-time duration can represent the small values of the radiation efficiency parameter (η_R) [19]. This suggests that the fracture energy of seismic deformation is much greater than the radiated energy. This simple relation can determine the dynamic character of the 2011 Van earthquake [27]. This study has shown that for the 2011 Van earthquake, the radiation efficiency, which is given by $\eta_R = 2\mu \tilde{e}/\Delta\sigma_s$, is smaller than 1. This value implies that the microscopic breakdown process dominates the dynamics of faulting and the fracture energy [19]. The amount of energy mechanically dissipated during rupture is relatively larger than that of the energy radiated as seismic waves. Time period between the 20th and 90th days (high b-value in **Figure 7**) can be interpreted as a fault or fracture density-dominated period. This period implies a maximum fracture energy and related breakdown micromechanisms occurring during the Van mainshock.

5.3. The shallow gas and microseismicity

Various shallow gas indicators have been observed from multichannel seismic reflection and chirp data from Lake Van [21]. Strong evidence of the shallow gas accumulations in Lake Van includes acoustic blanking, pockmarks, seismic chimneys, bright spots, mound-like features, and enhanced reflections (**Figures 9b** and **c**).

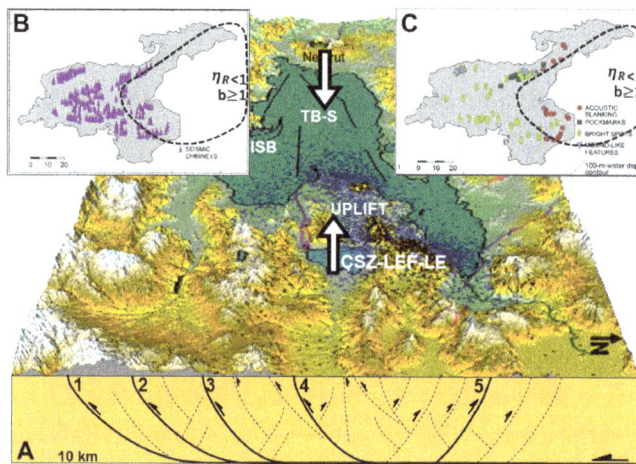

Figure 9. 3D-faulting and deformation map of Lake Van basin with aftershock seismicity. 3D-digital elevation block diagram of Lake Van basin with its boundary faults and epicentral distribution of the Van aftershock sequence show deformational relations between faults, deep Tatvan basin (TB), and aftershocks in A (see **Figure 3**). Inferred transpressional faults (dextral) at crustal depths compiled from previous studies [10, 29, 40] are hypothetically shown in this diagram. Numbers (1–5) represent faults. Number 1 is MS Mus Suture, 2 is GF Gevas Fault, 3 is EF Edremit Fault, 4 is F-KF Fault-Kalecik Fault, and 5 is eastern segment of NM-bf, Northern Margin Boundary Fault (see **Figure 3** for these faults). Distribution maps show seismic chimneys in B, acoustic blanking, pockmarks, bright spots, and mound-like features in C (data obtained from Ref. [21]). The wide area shown by black-dashed line indicates the epicentral image of aftershock hypocenters at focal depths between 5 and 25 km. The inverse relationship between the radiation efficiency and the b-value is compatible with depths at 5–25 km shown in B and C (see b-value depth profiles in Ref. [20]). TB-S, Tatvan Basin-Subsidence; ÇSZ-LEF-LE, Çarpanak Spur Zone-Lake Erçek Fault-Lake Erçek; and ISB, Internal Sub-Basin.

The enhanced reflections observed at more than 200 locations suggest the presence of free gas in Lake Van with seismic chimneys, suggesting the emission of gases (**Figure 9b**). Some of these chimneys have pockmarks (**Figure 9c**), which may be vertical vents or conduits for active fluid emission. Faults, fractures, and fissures may provide migration pathways for deep gas in TB. The acoustic blanking (**Figure 9c**) indicates that changes in the hydrostatic pressure may control the formation and preservation of the gas-charged zones. The mound-like features (mud volcanoes) (**Figure 9c**) suggest active gas emission and venting activity. Bright spots, which are a characteristic of gas accumulations, indicate gas-charged zones (**Figure 9c**). These pockmarks are only seen in the northeastern part of the lake. The absence of pockmarks in other parts of the lake (**Figure 9c**) can be due to a higher permeability of the lake sediments [14, 21].

These observations show that the thicker and unconsolidated soft sedimentary section, and the deformation style of Lake Van (turbiditic wedges, debris flows, tephra deposits, and progradational clinoform packages and the steep oblique-slip boundary faults) [12–14, 21]) have the potential to provide ideal conditions that allow the sediments in TB (**Figure 9a**) to act as a gas and/fluid reservoir (**Figures 9b** and **c**). This suggests that a gas-soft sediment mixture modulates the preferential emission of deep-sourced fluids through the faults, fractures, and/ or fissures in Lake Van and in particular aseismic TB (**Figure 9a**). The abrupt changes in the sedimentary section hamper the fluid transport in the sediments of TB. This process of the gas-sediment-fluid mixture in Lake Van has the potential of many smaller and same-sized magnitude aftershocks (microseismicity) that increases b-value ($b > 1.0$) and dramatically decreases the rate of seismic radiation ($\eta_R \ll 1$) (**Figures 9b, c**, and **10**). Since, the gas accumulations with the local release of deep and/or shallow fluids, soft sediment deformations, and fractures in Lake Van can amplify each other [24] and control the asymmetric distribution and localization of aftershock events as a function of depth (**Figure 10**).

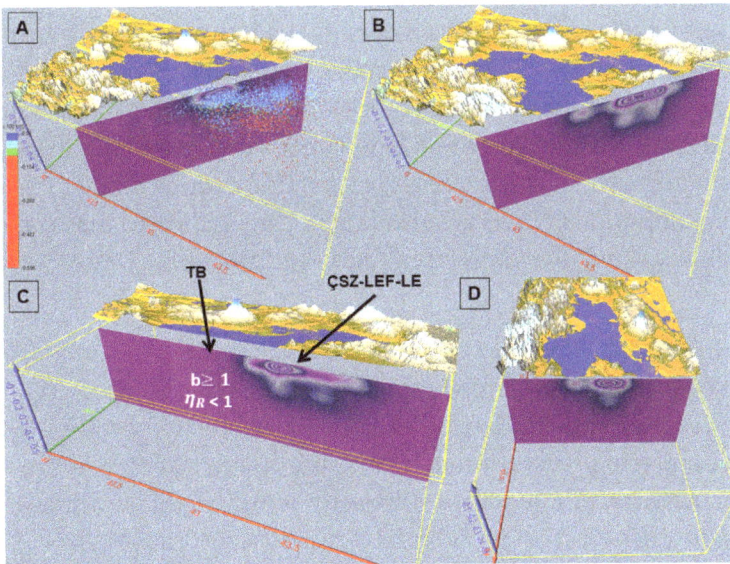

Figure 10. Cluster patterns of seismic density of aftershock events. High-resolution cross-sectional images of the hypocenter distribution of 5753 events indicate the localized cluster patterns of seismic density of events at crustal depths shown by sections projected on the shaded relief map. NE-SW projections are shown by A and B and W-E and N-S projections are shown by C and D, respectively. NE-SW cross section seen in A is the section toward the east to show event clusters (B) in different projections. Note that the deep Tatvan basin (TB) of Lake Van is aseismic, indicating no events recorded and the focal area is highly seismic with event clusters seen in B, C, and D, indicating the uplifted (ÇSZ-LEF-LE) zone (see **Figure 3**) (focal depths of 5753 events in km × 100 and given by colored scale in A).

The epicenters of aftershock events are densely clustered in a wedge-shaped narrow area, bounded by the west-east extending ÇM-bf, F-KF and NM-bf. This narrow area is a fault-bounded, uplifted block by ÇM-bf and NM-bf (ÇSZ-LEF-LE) (**Figures 3**, **9a**, and **10**). ÇSZ is interpreted to have been cut by an inferred LEF extending from LE in the east to Lake Van in the west (**Figures 3** and **9a**). The deformational features of Lake Van, especially ÇSZ, ÇM-bf, and NM-bf, are structurally related to the two mainshocks of the Van and Edremit events (**Figure 2a**) and the variability in the b-value. In this study, a concept of seismically active (ÇSZ-LEF-LE) and passive (TB) zones is proposed for the whole Lake Van region (**Figures 3**, **9a**, and **10**) and it is argued that the results in this study may indicate TB as a relaxation barrier and ÇSZ-LEF-LE as a fragmentation barrier [86] (**Figures 9b**, **c**, and **10**).

This study for the first time takes into account the low seismic radiation (high seismic deformation), considering the b-value combined with the radiation efficiency (η_R) that also represents microprocesses and differential stress in the rupture area.

6. Conclusions

In this study, the time and depth-dependent variability in the b-value of the G-R relation, the radiation efficiency (η_R), and related aftershock seismicity patterns were investigated for the Van earthquake of 23 October 2011. The main conclusions are summarized as follows:

The variation in b-value is associated with the two mainshocks in Van (23 October 2011) (M_w, 7.1) and Edremit (9 November 2011) (M_w, 5.6) earthquakes. We assume a possibility that the changes in b-value reflect changes in the local stress state in the vicinity of the aftershocks. The variability in the b-values may be correlated with decreasing and increasing effective stress levels near the focal area. The increase of b-value implies that there was an increase in material heterogeneity and a decrease in shear stress. The b-value drop is likely to infer an increase of effective stress and continuous stress accumulation. This indicates that there are potentially locked zones in and around the focal area (seismic ÇSZ-LEF-LE area as a fragmentation barrier).

The main result of the 2011 Van and Edremit earthquakes is that they considerably lowered the general shear stress in the area (high b-value) and increased complexity in and around the focal zone. The epicentral distribution of aftershock seismicity reveals a chaotic picture of clustering of events and irregular variation in magnitude as a function of time. The anomalous occurrence and distribution of aftershock seismicity with variable b-values over 170 days fits an inhomogeneously stressed crust with high fracture energy, causing low (high b-value) and high shear stress (low b-value). Differing amounts of rapid stress changes over 170 days or the type of driven mechanism could have caused the variation in the b-values. The high and low b-values over 170 days suggest that remarkable physical and mechanical changes occurred in the seismic deformation potential of the crust. The time-dependent seismic variability of the events with varying b-values for different time intervals is an indicator of a situation of stress transfer from a lower to a higher stress, associated with active tectonic deformation and the faulting style of the Lake Van basin.

The small radiation efficiency (η_R) (low seismic radiation/high seismic deformation), considering the high b-value combined with gas-fluid-fracture-sediment mixture in aseismic TB (relaxation barrier) represents high fracture energy and differential stress in the rupture area. The complexity of the Van earthquake rupture pattern suggests that the microscopic processes (high b-value) on the fault planes seem to have played very important role in controlling the rupture dynamics and deep basinal section (TB) of Lake Van. This indicates that more than a single process controlled local stress state and the rupture process in the 2011 Van earthquake. The slip zone was governed by the complex processes and this caused the occurrence of a significant rise (low b-value from the distinct slip localization of locked or strained faults) and drop (high b-value) in the friction near the focal area. This assumes that an amount of energy was not radiated and deposited near the focal region in the form of fracture and frictional energy in the 2011 Van event.

Finally, the seismic uniqueness of the 2011 Van earthquake must be considered. This research is the first to take into account the low seismic radiation (high seismic deformation) and examine the b-values associated with active tectonic deformation and faulting style of the epicentral region following the Van earthquake. In this study, as different from the previous studies, a first attempt has been made to interpret the variations in the b-value combined with the radiation efficiency (η_R) and the cross-sectional profiles of hypocenters of aftershocks as most essential seismic parameters of possible increases in the nonlinear stresses. This research has significant implications for the 2011 Van mainshock as an intraplate seismogenic paradigm and the thin-skinned seismic deformation of the Lake Van area during the post-collisional period.

Acknowledgements

The author thanks all members of the National Earthquake Monitoring Center (NEMC) at the Kandilli Observatory and Earthquake Research Institute (KOERI, Turkey) for the continuous seismological data. The author is grateful to Prof. Dr. Alper Çabuk for helping in using and processing the earthquake data and the seismological laboratory, Prof. Dr. G. Berkan Ecevitoğlu for providing the aftershocks monitoring FORTRAN code and commenting on concluding remarks of this study. The author offers sincere thanks to Leader of Lake Van Project seismic survey, Prof. Dr. Sebastian Krastel (Kiel, Germany) for providing multichannel seismic reflection profiles (International Continental Drilling Program, ICDP-PaleoVan Project-2004 funded by Germany Money Foundation, "Deutsche Forschungs Gemeinschaft-DFG") collected from Lake Van basin, Prof. Dr. A. M. Celal Şengör for commenting on faulting style and tectonics of the study area and Prof. Dr. Elena Kozlovskaya (Oulu, Finland) for providing the Laboratory of Applied Seismology (Seismic Handler Manual software), University of Oulu. Also, the author offers his greatest thanks to editors and the two anonymous reviewers for their constructive comments and suggestions which helped to improve the manuscript. Some figures are generated by Generic Mapping Tools (GMT) code developed by [89].

Author details

Mustafa Toker

Address all correspondence to: tokermu@gmail.com

Department of Geophysical Engineering, Yuzuncu Yıl University, Van, Turkey, Sodankylä
Geophysical Observatory (SGO), University of Oulu, Oulu, Finland

References

[1] Sengör AMC, Görür N, Saroğlu F. Strike-slip faulting and related basin formation in zones of tectonic escape. Turkey as a case study in Strike-slip faulting and basin formation. In: Biddle KT, Christie-Blick N, editors. Special Publication of Society of Economical Paleontology and Mineralogy. London, UK; 1985. p. 227-264.

[2] Dewey JF, Hempton MR, Kidd WSF, Saroğlu F, Sengör AMC. Shortening of continental lithosphere: the neotectonics of Eastern Anatolia—a young collision. In: Coward MP, Ries AC, editors. Collision Tectonics. Geological Society; London, UK; 1986. p. 3-36.

[3] McClusky S, Balassanian S, Barka A, Demir C, Ergintav S, Georgiev I, Gürkan O, Hamburger M, Hurst, K, Kahle H, Kastens K, Nadariya M, Ouzouni A, Paradissis D, Peter Y, Prilepin M, Reilinger R, Sanli I, Seeger H, Tealeb A, Toksöz MN, Veis G. GPS constraints on plate kinematics and dynamics in the Eastern Mediterranean and Caucasus. Journal of Geophysical Research. 2000;105:5695–5719.

[4] McKenzie DP. Active tectonics of the Mediterranean region. Geophysical Journal of Royal Astronomical Society. 1972;30:109-185.

[5] Toksöz MN, Nabelek J, Arpat E. Source properties of the 1976 earthquake in eastern Turkey: a comparison of field data and teleseismic results. Tectonophysics. 1978;49:199-205.

[6] Taymaz T, Eyidoğan H, Jackson J. Source parameters of large earthquakes in the East Anatolian fault zone (Turkey). Geophysical Journal International. 1991;106:537-550.

[7] Saroğlu F, Emre Ö, Kusçu İ. Active Fault Map of Turkey. Ankara, Turkey: Mineral Research and Exploration Institute of Turkey; 1992.

[8] Pınar A. Rupture process and spectra of some major Turkish earthquakes and their seismotectonic implications [dissertation]. Istanbul: Boğaziçi Universitesi; 1995.

[9] Utkucu M. 23 October 2011 Van, Eastern Anatolia, earthquake (Mw 7.1) and seismotectonics of Lake Van Area. Journal of Seismology. 2013;7(5):23-33.

[10] Utkucu M. Implications for the water-level-change triggered moderate (M≥4.0) earthquakes in Lake Van basin, Eastern Turkey. Journal of Seismology. 2006;10:105-117. DOI10.1007/s10950-005-9002

[11] Utkucu M, Durmus H, Yalçın H, Budakoğlu E, Isık E. Coulomb static stress changes before and after the 23 October 2011 Van, eastern Turkey, earthquake (Mw = 7.1): Implications for the earthquake hazard mitigation. Natural Hazards and Earth System Science. 2013;13:1-14. DOI: 10.5194/nhess-13-1-2013

[12] Toker M. Tectonic and magmatic structure of Lake Van basin and its structural evolution, Eastern Anatolia Accretionary Complex (EAAC), E-Turkey [thesis]. Istanbul: Istanbul Technical University; 2011.

[13] Toker M, Sengör AMC. Structural Elements, Tectonic and Sedimentary Evolution of Lake Van Basin, Eastern Turkey, Journal of Istanbul Technical University (ITU)/D, Engineering. 2011; 10 (4):119-130.

[14] Çukur D, Krastel S, Demirel-Schlüter F, Demirbağ E, Imren C, Nissen F, Toker M, PaleoVan-Working Group. Sedimentary evolution of Lake Van (Eastern Turkey) reconstructed from high-resolution seismic investigations. International Journal of Earth Sciences (Geol Rundsch). 2013;102(2):571-585. DOI: 10.1007/s00531-012-0816-x

[15] Irmak TS, Doğan B, Karakas A. Source mechanism of the 23 October 603 2011 Van (Turkey) earthquake (Mw = 7.1) and aftershocks with its tectonic implications. Earth, Planets and Space. 2012;64 1-13.

[16] Toker M. Time-dependent analysis of aftershock events and structural impacts on intraplate crustal seismicity of the Van earthquake (Mw 7.1, 23 October 2011), E-Anatolia. Central European Journal of Geosciences. 2013 5(3) 423-434. DOI: 10.2478/s13533-012-0141-8

[17] Fielding EJ, Lundgren PR, Taymaz T, Yolsal-Çevikbilen S, Owen SE. Fault-slip source models for the 2011 M 7.1 Van Earthquake in Turkey from SAR Interferometry, Pixel Offset Tracking, GPS, and seismic waveform analysis. Seismological Research Letters. 2013;84(4):579-593. DOI: 10.1785/0220120164

[18] Husseini M. Energy balance for motion along a fault. Geophysical Journal Royal of Astronomical Society. 1977;49:699-714.

[19] Kanamori H, Brodsky EE, Institute of Physics Publishing, Reports on Progress in Physics (Rep. Prog. Phys.) 67 (2004) page range: 1429–1496. PII: S0034-4885(04)25227-7. doi:10.1088/0034-4885/67/8/R03, © 2004 IOP Publishing Ltd. Printed in the UK.

[20] Bayrak Y, Yadav RBS, Kalafat D, Tsapanos TM, Çınar H, Singh AP, Bayrak E, Yılmaz S, Öcal F, Koravos G. Seismogenesis and earthquake triggering during the Van (Turkey) 2011 seismic sequence. Tectonophysics. 2013;601:163-176. DOI: 10.1016/j.tecto.2013.05.008

[21] Çukur D, Krastel S, Tomonaga Y, Çagatay N, Meydan AF, the PaleoVan Science Team. Seismic evidence of shallow gas from Lake Van, eastern Turkey. Marine and Petroleum Geology. 2013;48:341-353. DOI: 10.1016/j.marpetgeo.2013.08.017

[22] Khilyuk LF, Chilinger GV, Enders B, Robertson Jr JO. Gas Migration: Events Preceding Earthquakes. USA: Gulf Publishing Company Press; 2000.

[23] Gurevich AE, Chilingarian GV. Subsidence over producing oil and gas fields, and gas leakage to the surface. Journal of Petroleum Science Engineering. 1993;9:239-250.

[24] Katz SA, Khilyuk LF, Chilingarian GV. Interrelationships among subsidence owing to fluid withdrawal, gas migration to the surface and seismic activity: Environmental aspects of oil production in seismically active areas. Journal of Petroleum Science Engineering. 1994;11:103-112.

[25] Litt T, Krastel S, Sturm M, Kipfer R, Örçen S, Heumann G, Franz SO, Ulgen UB, Niessen F. 'PALEOVAN', International Continental Scientific Drilling Program (ICDP): site survey results and perspectives. Quaternary Science Reviews. 2009;28:1555-1567. DOI: 10.1016/j.quascirev.2009.03.002

[26] Toker M. Time-dependent aftershock seismicity patterns and their propagation-the Van earthquake (7.2), 23 Oct. 2011, E-Anatolia. In: Proceedings of the Seventh Congress and Technical Exhibition of Balkan Geophysical Society; 7-10 October 2013; Tirana, Albania. 2013.

[27] Toker M. Discrete characteristics of the aftershock sequence of the 2011 Van earthquake. Journal of Asian Earth Sciences. 2014;92:168-186. DOI: 10.1016/j.jseaes.2014.06.015

[28] Horasan G, Boztepe-Güney A. Observation and analysis of low-frequency crustal earthquakes in Lake Van and its vicinity, eastern Turkey. Journal of Seismology. 2006;DOI: 10.1007/s10950-006-9022-2

[29] Sengör AMC, Özeren MS, Keskin M, Sakınç M, Özbakır AD, Kayan İ. Eastern Turkish high plateau as a small Turkic-type orogen: Implications for post-collisional crust-forming processes in Turkic-type orogens. Earth-Science Reviews. 2008;90(1-2):1-48.

[30] Degens ET, Kurtman F. The Geology of Lake Van. Ankara, Turkey: MTA Press; 1978. 158 p.

[31] Kipfer R, Aeschbach-Hertig W, Baur H, Hofer M, Imboden DM, Signer P. Injection of mantle type Helium into Lake Van (Turkey): The clue for quantifying deep water renewal. Earth and Planetary Science Letters. 1994;125:357-370.

[32] Keskin M. Magma generation by slab steepening and break off beneath a subduction-accretion complex. An alternative model for collision related volcanism in eastern Anatolia. Geophysical Research Letters. 2003;30:8046. DOI: 10.1029/2003GL018019

[33] Sengör AMC, Özeren MS, Genç C, Zor E. East Anatolian high plateau as a mantle-supported, north-south shortened domal structure. Geophysical Research Letters. 2003;30(24):8045. DOI: 10.1029/2003GL017858

[34] Litt T, Anselmetti FS, Cagatay MN, Kipfer R, Krastel S, Schmincke HU, PaleoVan Working Team. A 500,000 year-long sedimentary archive drilled in Eastern Anatolia (Turkey), The PaleoVan Drilling Project. EOS. 2011;92(51):453-464.

[35] Litt T, Anselmetti FS, Baumgarten H, Beer J, Cagatay N,Çukur D, Damci E, Glombitza C, Heumann G, Kallmeyer J, Kipfer R, Krastel S, Kwiecien O, Meydan AF, Orcen S, Pickarski N, Randlett M-E, Schmincke H-U, Schubert CJ, Sturm M, Sumita M, Stockhecke M, Tomonaga Y, Vigliotti L, Wonik T, the PALEOVAN Scientific Team. 500,000 years of environmental history in Eastern Anatolia: the PALEOVAN Drilling Project of Scientific Drilling. 2012;14:18-29.

[36] Wong HK, Finckh P. Shallow structure in Lake Van, eastern Turkey. In: Degens ET, Kurtman F, editors. The Geology of Lake Van. Ankara, Turkey: MTA Press; 1978. p. 20-28.

[37] Wong HK, Degens ET. The bathymetry of Lake Van, eastern Turkey. In: Degens ET, Kurtman F, editors. The Geology of Lake Van. Ankara, Turkey: MTA Press; 1978. p. 6-10.

[38] Degens ET, Wong HK, Kurtman F, Finkch P. Geological development of Lake Van: A summary. In: Degens ET, Kurtman F, editors. The Geology of Lake Van. Ankara, Turkey: MTA Press; 1978. p. 134-146.

[39] Emre Ö, Duman TY, Özalp S, Elmacı H. The Preliminary Reports on Field Survey Observations of the 23th October 2011 Van Earthquake and Ruptured Fault. Ankara, Turkey: MTA Department of Geological Researches; 2011. 22 pp.

[40] Ketin I. A short explanation about the results of observations made in the region between Lake Van and Iranian border. Bulletin of Geological Society of Turkey. 1977;20:79-85.

[41] Lahn E. A note about earthquakes in Van area (July-November 1945). Maden Tetkik ve Arama Enstitüsü. 1946;35:126-132.

[42] Tasman CE. Varto and Van earthquakes. Publication of Mineral Resources and Exploration Institute of Turkey. 1946;2(6):287-291.

[43] Özkaymak Ç. Active Tectonic Features of Van city and its surroundings [dissertation]. Van, Turkey: Yuzuncu Yıl University; 2003.

[44] Özkaymak Ç, Yürür T, Köse O. An example of intercontinental active collisional tectonics in the Eastern Mediterranean region (Van, eastern Turkey). In: Proceedings of the 5th International Symposium on Eastern Mediterranean Geology; 14-20 April 2004; Thessaloniki, Greece. 2004. p. 591-593.

[45] Koçyiğit A. Neotectonics and seismicity of East Anatolia. In: Van, Turkey. Yüzüncü Yıl University; 2002. p. 2-4.

[46] Özler M. Karst hydrogeology of Kusluk-Dilmetas karst springs, Van-Eastern Turkey. Environmental Geology. 2001;41:257-268.

[47] Miller AD, Steward RC, White RA, Luckett R, Baptie BJ, Aspinall WP, Latchman JL, Lynch LL, Voight B. Seismicity associated with dome grows and collapse at Soufriére Hills volcano, Montserrat. Geophysical Research Letters. 1998;25:3401-3404.

[48] Mogi K. Study of elastic shocks caused by the fracture of heterogeneous materials and its relation to earthquake phenomena. Bulletin of Earthquake Research Institute University of Tokyo. 1962;40:125-173.

[49] Warren NM, Latham GV. An experiment study of thermal induced microfracturing and its relation to volcanic seismicity. Journal of Geophysical Research. 1970;75:4455-4464.

[50] Boğaziçi University, Kandilli Observatory and Earthquake Research Institute. KOERI [Internet]. 2011–2012. Available from: http://www.koeri.boun.edu.tr.

[51] Gülen L, Pınar A, Kalafat D, Özel N, Horasan G, Yılmazer M, Isıkara AM. Surface fault breaks, aftershocks distribution, and rupture process of the 17 August-1999 Izmit, Turkey, earthquake. Bulletin of the Seismological Society of America. 2002;92:230-244.

[52] Gutenberg B, Richter CF. Frequency of earthquakes in California. Frequency of earthquakes in California. Bulletin Seismological Society of America. 1944;34:185-188.

[53] Scholz CH. The frequency–magnitude relation of microfracturing in rock and its relation to earthquakes. Bulletin Seismological Society of America. 1968;58:399-415.

[54] Wyss M. Toward a physical understanding of earthquake frequency distribution. Geophysical Journal of Royal Astronomical Society. 1973;31:341-359.

[55] Amitrano D. Brittle-ductile transition and associated seismicity: Experimental and numerical studies and relationship with the b value. Journal of Geophysical Research. 2003;108(B1)DOI: 10.1029/2001JB000680

[56] Schorlemmer D, Wiemer S, Wyss M. Variations in earthquake-size distribution across different stress regimes. Nature. 2005;437:539-542. DOI: 10.1038/nature04094

[57] Frohlich C, Davis SD. Teleseismic b values or much ado about 1.0. Journal of Geophysical Research. 1993;98:631-644. DOI: 10.1029/92JB01891

[58] Utsu T. Seismicity Studies: A Comprehensive Review. Japan: University of Tokyo Press; 1999. 876 p.

[59] Utsu T. Relationships between magnitude scales. In: Lee WHK, editor. International Handbook of Earthquake and Engineering Seismology, Part A. Amsterdam: Academic Press. 2002. p. 733-745.

[60] Utsu T. Statistical features of seismology. In: Lee WHK, editor. International Handbook of Earthquake and Engineering Seismology, Part A. Amsterdam: Academic Press. 2002. p. 719-732.

[61]   Wiemer S, Wyss M. Spatial and temporal variability of the b-value in seismogenic volumes: an overview. Advances in Geophysics. 2002;45:259-302.

[62]   Yadav RBS, Papadimitriou EE, Karakostas VG, Shanker D, Rastogi BK, Chopra S, Singh AP, Santosh K. The 2007 Talala, Saurashtra, western India earthquake sequence: tectonic implications and seismicity triggering. Journal of Asian Earth Sciences. 2011;40:303-314.

[63]   Yadav RBS, Gahalaut VK, Chopra S, Shan B. Tectonic implications and seismicity triggering during the 2008 Baluchistan, Pakistan earthquake sequence. Journal of Asian Earth Sciences. 2012;45:167-178.

[64]   Bridges DL, Gao SS. Spatial variation of seismic b-values beneath Makushin Volcano, Unalaska Island, Alaska. Earth Planetary Science Letters. 2006;245:408-415.

[65]   Zamani A, Agh-Atabai M. Multifractal analysis of the spatial distribution of earthquake epicenters in the Zagros and Alborz-Kopeh Dagh regions of Iran. Journal of Science and Technology. 2011;A1:39-51.

[66]   Sorbi MR, Nilfouroushan F, Zamani A. Seismicity patterns associated with the September 10th, 2008 Qeshm earthquake, South Iran. International Journal of Earth Sciences (Geol Rundsch). 2012; 12 45-57 DOI: 10.1007/s00531-012-0771-6

[67]   Utsu T. A method for determining the value of b in a formula logN = a − bM showing the magnitude frequency for earthquakes. Geophysical Bulletin of Hokkaido University. 1965;13:99-103.

[68]   Wiemer S. A software package to analyze seismicity: ZMAP. Seismological Research Letters. 2001;72(3):373-382.

[69]   Aki K. Maximum likelihood estimate of b in the formula log N = a-bM and its confidence limits. Bulletin of Earthquake Research Institute. 1965;43:237-239.

[70]   Stein S, Wysession M. An Introduction to Seismology: Earthquakes and Earth Structure. UK: Blackwell Publishing Press; 2002.

[71]   Shi Y, Bolt B. The standard error of the magnitude–frequency b value. Bulletin of the Seismological Society of America. 1982;72(5):1677-1687.

[72]   Wiemer S, Wyss M. Minimum magnitude of completeness in earthquake catalogs: examples from Alaska, the western United States, and Japan. Bulletin of Seismological Society of America. 2000;90(4):859-869. DOI: 10.1785/0119990114

[73]   Rydelek PA, Sacks IS. Testing the completeness of earthquake catalogs and the hypothesis of self-similarity. Nature. 1989;337:251-253.

[74]   Aki K. Generation and propagation of G waves from the Niigata earthquake of June 16, 1964: Part 1: A statistical analysis. Bulletin of Earthquake Research Institute of Tokyo University. 1966;44:73-88.

[75]   Wyss M, Brune JN. Seismic moment, stress, and source dimensions for earthquakes in the California-Nevada region. Journal of Geophysical Research. 1968;73:4681-4694. DOI: 10.1029/JB073i014p04681

[76] Choy GL, Boatwright JL. Global patterns of radiated seismic energy and apparent stress. Journal of Geophysical Research. 1995;100:18205-18228. DOI: 10.1029/95JB01969

[77] Venkataraman A, Kanamori H. Observational constraints on the fracture energy of subduction zone earthquakes. Journal of Geophysical Research. 2004;109(B5):302. DOI: 10.1029/2003JB002549

[78] Ide S, Beroza GC. Does apparent stress vary with earthquake size. Geophysical Research Letters. 2001;28:3349-3352. DOI: 10.1029/2001GL013106

[79] Scholz CH. The Mechanics of Earthquakes and Faulting. London, England: Cambridge University Press; 1990. 439 p.

[80] Ben-Zion Y. Collective behavior of earthquakes and faults: Continuum-discrete transitions, progressive evolutionary changes, and different dynamic regimes. Reviews of Geophysics. 2008;46:RG4006. DOI: 10.1029/2008RG000260

[81] Gorgun E, Zang A, Bohnhoff M, Milkereit C, Dresen G. Analysis of Izmit aftershocks 25 days before the November 12th 1999 Düzce earthquake, Turkey. Tectonophysics. 2009;474:507-515.

[82] Tsukakoshi Y, Shimazaki K. Decreased b-value prior to the M 6.2 Northern Miyagi, Japan, earthquake of 26 July 2003. Earth, Planets and Space. 2008;60:915-924.

[83] Wiemer S, Katsumata K. Spatial variability of seismicity parameters in aftershock zones. Journal of Geophysical Research. 1999;104:13135-13151.

[84] Goria PD, Akinci A, Lucentea FP, Kılıç T. Spatial and temporal variations of aftershock activity of the 23 October 2011 Mw 7.1 Van, Turkey earthquake. Bulletin of the Seismological Society of America. 2014;104(2):913-930. DOI: 10.1785/0120130118

[85] Nilsen TH, Sylvester AG. Strike-slip basins. In: Busby CJ, Ingersoll RV, editors. Tectonics of Sedimentary Basins. Oxford: Blackwell Scientific Publications; 1995. p. 425-457.

[86] King GCP. Speculations on the geometry of the initiation and termination processes of earthquake rupture and its relation to morphology and geological structure. Pageography. 1986;124(3):567-586.

[87] Tolstoy M, Bohnenstiehl DM, Edwards M, Kurras G. The seismic character of volcanic activity at the ultra-slow spreading Gakkel Ridge. Geology. 2001;29:1139-1142.

[88] Akinci A, Antonioli A. Observations and stochastic modelling of strong ground motions for the 2011 October 23 Mw 7.1 Van (Turkey) earthquake. Geophysical Journal International. 2013;192(3):1217-1239. DOI: 10.1093/gji/ggs075

[89] Wessel P, Smith WHF. New, improved version of the generic mapping tools released. EOS. 1998;79(47):579.

5

Application of Local Wave Decomposition in Seismic Signal Processing

Ya-juan Xue, Jun-xing Cao, Gu-lan Zhang,
Hao-kun Du, Zhan Wen, Xiao-hui Zeng and
Feng Zou

Additional information is available at the end of the chapter

Abstract

Local wave decomposition (LWD) method plays an important role in seismic signal processing for its superiority in significantly revealing the frequency content of a seismic signal changes with time variation. The LWD method is an effective way to decompose a seismic signal into several individual components. Each component represents a harmonic signal localized in time, with slowly varying amplitudes and frequencies, potentially highlighting different geologic and stratigraphic information. Empirical mode decomposition (EMD), the synchrosqueezing transform (SST), and variational mode decomposition (VMD) are three typical LWD methods. We mainly study the application of the LWD method especially EMD, SST, and VMD in seismic signal processing including seismic signal de-noising, edge detection of seismic images, and recovery of the target reflection near coal seams.

Keywords: local wave decomposition, empirical mode decomposition, the synchrosqueezing transform, variational mode decomposition, seismic signal processing

1. Introduction

The main characteristics of a nonlinear and nonstationary signal are that the signal is changed with the time and its frequency is transient and only in the presence of a local time, that is, a local wave signal. The conventional Fourier transform is only suitable for the global wave, and it shows great limitation for the analysis of the local wave. The thereafter appeared time-frequency methods including short-time Fourier transform, wavelet transform, Wigner-Ville distribution, and so on can appropriately describe the time variation of the nonstationary signal to some extent and show advantages over the Fourier transform. But overall, they are still in the scope of global wave.

Local wave decomposition (LWD) is a class of local wave processing methods that suit for the analysis of the nonlinear and nonstationary signals. LWD can decompose a complex multi-frequency component signal into a number of monofrequency or narrow-band signals, that is, intrinsic mode function (IMF), which has the better Hilbert transform characteristics and the calculation of its instantaneous frequency makes sense. LWD has evolved from empirical mode decomposition (EMD), which is first introduced by Huang et al. [1], to the recent works including ensemble empirical mode decomposition (EEMD) [2], complete ensemble empirical mode decomposition (CEEMD) [3], the synchrosqueezing transform (SST) [4], variational mode decomposition [5], and so on. Among these LWD methods, here we mainly study three typical methods: EMD, SST, and VMD. EMD is the method, which has already been widely used in time-frequency applications in various signal processing such as climate analysis [6], seismic signal processing [7], mechanical fault diagnosis [8], speech signal processing [9], medicine and biology signal processing [10, 11], and so on. Due to the lack of mathematical understanding and some other obvious shortcomings including end effects [12], the problem of stopping criterion [13, 14], the influence of sampling [15], spline problems and mode mixing [16, 17], and so on in EMD algorithm, the other tools and improved methods pursuing the same goal are developed. SST is a wavelet-based EMD-like tool which has a firm theoretical foundation [18]. It introduces precise mathematical definition of a class of functions construct-ed by several limited approximate harmonic components. SST is a combination of wavelet analysis and frequency reassignment. While VMD is another newly adaptive and fully intrinsic method that has a firm theoretical foundation. It decomposes a nonlinear and nonstationary signal with quasi-orthogonal, nonrecursive manner into a series of band-limited subsignals.

Since a seismic signal is a nonlinear and nonstationary signal, the LWD method is more suitable for the analysis of the seismic signal than the other traditional Fourier- and wavelet-based methods. Each obtained IMF with different narrow frequency band from LWD method represents a harmonic signal localized in time, with slowly varying amplitudes and frequen-cies, potentially highlighting different geologic and stratigraphic information and their pore fluids.

In this chapter, we mainly study the application of the LWD method especially three typical LWD methods, that is, EMD, SST, and VMD, in seismic signal processing including seismic signal de-noising, edge detection of seismic images, and recovery of the target reflection near coal seams.

2. The general theory of local wave decomposition

In this chapter, not all the LWD methods that might be useful in practice can be examined. Guided by the various previous literatures, we only briefly review a few concepts and tools on the most commonly used and three new typical LWD methods: EMD, SST, and VMD.

2.1. Empirical mode decomposition

The purpose of EMD is to obtain IMFs. Huang et al. [1] believe that any data consists of different simple intrinsic modes of oscillations, that is, IMFs. Each IMF involves only one mode of oscillation due to the forbiddance of no complex riding waves. With respect to the "local mean," the oscillation will also be symmetric. A meaningful instantaneous frequency of a multicomponent signal can be obtained by reducing the original signal to a collection of IMFs by employing EMD through a sifting process [19]. Only the instantaneous frequency of an IMF has physical meaning [1, 19]. An IMF is defined as a signal whose number of extrema and zero-crossings must either equal or differ at most by one and the mean value of the upper and lower envelopes, respectively, defined by the local maxima and minima is zero [1]. The IMF definition guarantees the narrow-band requirement for a stationary Gaussian process and the unwanted fluctuations induced by a symmetric waveform will not be emerged in the instantaneous frequency [1, 20]. The IMF is a linear or nonlinear signal that has constant amplitude and frequency as in a simple harmonic component or variable amplitude and frequency as functions of time.

EMD mainly includes the following steps:

1. Compute the average curve $m(t)$ of the original signal $x(t)$ for the upper envelope $u(t)$ and the lower envelope $v(t)$, which are respectively fitted by all the maxima and minima of $x(t)$:

$$m(t) = [u(t) + v(t)]/2, \tag{1}$$

in which, $v(t) \leq x(t) \leq u(t)$.

2. Subtract $m(t)$ from the original signal $x(t)$ and let $h_1(t) = x(t) - m(t)$.

3. Return to step (a) and replace $x(t)$ with $h_1(t)$. For $h_1(t)$, the corresponded upper and lower envelopes are $u_1(t)$ and $v_1(t)$, respectively. Repeat the above process until the resultant $h_k(t)$ meets the IMF definition:

$$m_1(t) = [u_1(t) + v_1(t)]/2$$

$$h_2(t) = h_1(t) - m_1(t)$$

$$.$$
$$.$$
$$.$$

$$m_{k-1}(t) = [u_{k-1}(t) + v_{k-1}(t)]/2$$

$$h_k(t) = h_{k-1}(t) - m_{k-1}(t).$$

Then the first IMF $c_1(t) = h_k(t)$. The residual part of the signal is: $r_1(t) = x(t) - c_1(t)$.

4. When there are more than one extremes (neither the constant nor the trend term) in $x(t)$, the EMD process is conducted until the remaining portion of the resulting signal is a monotone or a value less than a predetermined given value.

After the EMD process, the original signal $x(t)$ can be expressed as the sum of the IMFs and the margin:

$$x(t) = c_1(t) + c_2(t) + \cdots + c_n(t) + r_n(t) = \sum_{i=1}^{n} \text{Re}[a_i(t)\exp(j\theta_i(t))] + r_n(t) \tag{2}$$

where $c_i(t)$ is the ith IMF($i = 1 \sim n$)and $r_n(t)$ is the residue. $a_i(t)$ and $\theta_i(t)$ are respectively the instantaneous amplitude and phase of the ith IMF $c_i(t)$.

As shown above, the local mode of the signal is successively isolated from high frequency to low frequency based on the characteristic time scales by the sifting process in EMD. It is proved that EMD acts as an adaptive, multiband overlapping filter bank [21].

2.2. The synchrosqueezing transform

SST algorithm is another technique that aims to decompose a multicomponent signal into a series of IMFs like EMD. But unlike EMD, it introduces a precise mathematical definition for IMFs that can be viewed as a superposition of a reasonably small number of approximate harmonic components [4]. SST method uses frequency reassignment to improve the readability of wavelet-based time-frequency analysis.

The continuous wavelet transform (CWT) of a signal $s(t)$ is [22]:

$$W_s(a,b) = \frac{1}{\sqrt{a}} \int s(t)\psi^*(\frac{t-b}{a})dt \tag{3}$$

where a is the scale factor, b is the time shift factor, and ψ^* is the complex conjugate of the mother wavelet. The coefficients $W_s(a,b)$ which represent a concentrated time-frequency picture can be used to extract the instantaneous frequencies [22].

Rewrite Eq. (3) using Plancherel's theorem which states that the energy in the time domain equals energy in the frequency domain:

$$W_s(a,b) = \frac{1}{2\pi} \int \frac{1}{\sqrt{a}} \hat{s}(\xi) \hat{\psi}^*(a\xi) e^{jb\xi} d\xi \tag{4}$$

where ξ is the angular frequency. $\hat{\psi}(\xi)$ and $\hat{s}(\xi)$ are respectively the Fourier transform of $\psi(t)$ and $s(t)$. $j = \sqrt{-1}$.

Considering a single harmonic signal with the form:

$$s(t) = A\cos(\omega t) \tag{5}$$

Its Fourier pair can be expressed as

$$\hat{s}(\xi) = \pi A[\delta(\xi - \omega) + \delta(\xi + \omega)] \tag{6}$$

Then Eq. (4) can be transformed into

$$W_s(a,b) = \frac{A}{2} \int \frac{1}{\sqrt{a}} [\delta(\xi - \omega) + \delta(\xi + \omega)] \hat{\psi}^*(a\xi) e^{jb\xi} d\xi$$
$$= \frac{A}{2\sqrt{a}} \hat{\psi}^*(a\omega) e^{jb\omega} \tag{7}$$

Since the wavelet $\hat{\psi}^*(\xi)$ is condensed around its central frequency ω_0, $W_s(a,b)$ will be condensed around the horizontal line $a = \omega_0/\omega$, which represents the ratio of the central frequency of the wavelet to the central frequency of the signal. But in fact $W_s(a,b)$ always spreads out around this horizontal line, which makes a blurred projection in time-scale representation. This main smearing occurs in the scale dimension along the time shift factor b. And little smearing in dimension b along the scale axis can be found. It is proved that the instantaneous frequency $\omega_s(a,b)$ can be computed using the following equation when the smear in the b-dimension can be neglected:

$$\omega_s(a,b) = \frac{-j}{W_s(a,b)} \frac{\partial W_s(a,b)}{\partial b} \tag{8}$$

where for any point (a,b), $W_s(a,b) \neq 0$.

Then map the information from the time-scale plane to the time-frequency plane and convert every point (b, a) to $(b, \omega_s(a, b))$, which is named as synchrosqueezing operation. Since a and

b are discrete values, a scaling step $\Delta a_k = a_{k-1} - a_k$ can be used for any $a_{k'}$ where $W_s(a, b)$ is computed. Then the SST $T_s(w, b)$ is determined only at the centers ω_l with the frequency range $[\omega_l - \Delta\omega/2, \omega_l + \Delta\omega/2]$:

$$T_s(\omega_l, b) = \frac{1}{\Delta\omega} \sum_{a_k : |\omega(a_k, b) - \omega_l| \leq \Delta\omega/2} W_s(a_k, b) a^{-3/2} \Delta a_k \tag{9}$$

where $\Delta\omega = \omega_l - \omega_{l-1}$.

Eq. (9) shows that the new time-frequency representation of the signal $T_s(w, b)$ is synchrosqueezed along the frequency axis only [23]. The coefficients of the CWT are reallocated by SST to get a concentrated image over the time-frequency plane. The instantaneous frequency is also extracted from the new time-frequency representation [24].

The individual components s_k can be reconstructed by the discrete SST $\tilde{T}_{\tilde{s}}$:

$$s_k(t_m) = 2C_\varphi^{-1} \operatorname{Re}\left(\sum_{l \in L_k(t_m)} \tilde{T}_{\tilde{s}}(w_l, t_m) \right) \tag{10}$$

where C_φ is a constant dependent on the selected wavelet. Re(•) takes the real part of the representation. $\tilde{T}_{\tilde{s}}$ is the discretized version of $T_s(\omega_l, b)$, that is, $\tilde{T}_{\tilde{s}}(w_l, t_m)$. t_m is the discrete time, $t_m = t_0 + m\Delta t$, $m = 0, 1, ..., n - 1$. n is the total number of samples in the discrete signal $\tilde{s}_m$. Δt is the sampling rate.

2.3. Variational mode decomposition

VMD is a newly developed and theoretically well-founded EMD-like tool. It can nonrecursively decompose a multicomponent signal into a series of band-limited IMFs. Note here, IMF is defined more restrictively in VMD compared with the original IMF definition in EMD as the following [5]:

$$u(t) = A(t) \cos\big(\varphi(t)\big) \tag{11}$$

where $u(t)$ resembles an IMF which is a narrow-band amplitude-modulated, frequency-modulated signal. The phase $\varphi(t)$ is a nondecreasing function. The instantaneous frequency $\omega(t) = \varphi'(t)$ and the envelope $A(t)$ are nonnegative. $\omega(t)$ and $A(t)$ vary much slower than $\varphi(t)$ itself.

VMD is a constrained variational problem as shown by Eq. (12) [5, 25]:

$$\min_{\{u_k\},\{\omega_k\}} \left\{ \sum_k \left\| \partial_t \left[\left(\delta(t) + \frac{j}{\pi t} \right) * u_k(t) \right] e^{-j\omega_k t} \right\|_2^2 \right\} \tag{12}$$

subject to $\sum_k u_k = f$,

where u_k and ω_k are respectively the kth mode and their center frequency. For each mode, its bandwidth is estimated through the H^1 Gaussian smoothness of the demodulated signal. To address Eq. (12), a quadratic penalty and Lagrangian multipliers are used. The augmented Lagrangian is introduced as follows:

$$L(u_k, \omega_k, \lambda) = \alpha \sum_k \left\| \partial_t \left[\left(\delta(t) + \frac{j}{\pi t} \right) * u_k(t) \right] e^{-j\omega_k t} \right\|_2^2 \\ + \left\| f - \sum_k u_k \right\|_2^2 + \left\langle \lambda, f - \sum_k u_k \right\rangle \tag{13}$$

where α represents the balancing parameter of the data-fidelity constraint.

The main steps of VMD include the following:

1. Modes u_k update. Eq. (14) shows how the modes $\hat{u}_k^{n+1}(\omega)$ are updated:

$$\hat{u}_k^{n+1}(\omega) = \frac{\hat{f}(\omega) - \sum_{i<k} \hat{u}_i^{n+1}(\omega) - \sum_{i>k} \hat{u}_i^n(\omega) + \left(\hat{\lambda}^n(\omega)/2 \right)}{1 + 2\alpha \left(\omega - \omega_k^n \right)^2} \tag{14}$$

Eq. (14) clearly demonstrates that the modes $\hat{u}_k^{n+1}(\omega)$ acts as a Wiener filtering of the current residual with signal prior $1/(\omega - \omega_k^n)^2$. The mode in time domain can be obtained by taking the real part of the inverse Fourier transform of this filtered analytic signal.

2. Center frequencies ω_k^{n+1} update. The new center frequencies ω_k^{n+1} are put at the center of gravity of the corresponding mode's power spectrum which are shown in Eq. (15):

$$\omega_k^{n+1} = \frac{\int_0^\infty \omega \left| \hat{u}_k^{n+1}(\omega) \right|^2 d\omega}{\int_0^\infty \left| \hat{u}_k^{n+1}(\omega) \right|^2 d\omega} \tag{15}$$

3.

Dual ascent update. The Lagrangian multiplier $\hat{\lambda}^{n+1}$ is used as a dual ascent to enforce exact signal reconstruction. For all $\omega \geq 0$, $\hat{\lambda}^{n+1}$ is updated as shown in Eq. (16) until convergence $\sum_k \left\| \hat{u}_k^{n+1} - \hat{u}_k^n \right\|_2^2 / \left\| \hat{u}_k^n \right\|_2^2 < \varepsilon$

$$\hat{\lambda}^{n+1} = \hat{\lambda}^n + \tau \left(\hat{f} - \sum_k \hat{u}_k^{n+1} \right) \tag{16}$$

The detailed complete algorithm of VMD can be found in [5].

3. Applications

3.1. Synthetic data

In this section, synthetic data is first used to compare EMD with the SST and VMD methods. Then, the equivalent band-limited filter behavior of EMD, SST, and VMD are respectively investigated. Finally, to evaluate the noise robustness of the three methods, the synthetic data with added noise is analyzed.

The synthetic signal is shown in **Figure 1**. It is comprised of an initial 20 Hz cosine wave in which 90 Hz Ricker wavelets at 0.2 s, 75 Hz Ricker wavelets at 0.4 s, and two 40 Hz Ricker wavelets at 0.68 and 0.72 s are respectively superposed.

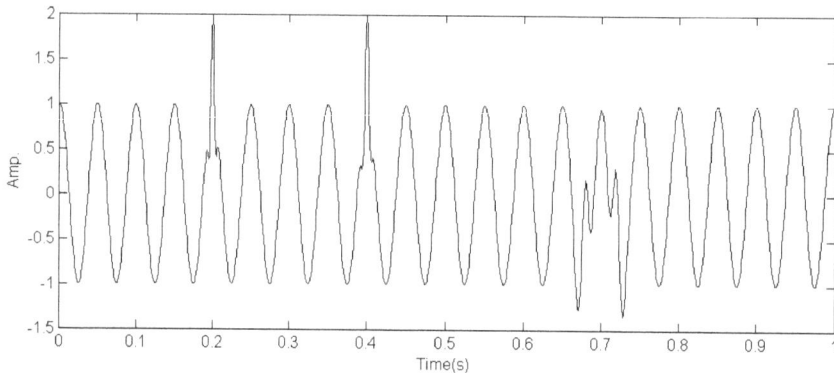

Figure 1. The synthetic signal. It is comprised of an initial 20 Hz cosine wave in which 90 Hz Ricker wavelets at 0.2 s, 75 Hz Ricker wavelets at 0.4 s, and two 40 Hz Ricker wavelets at 0.68 and 0.72 s are, respectively, superposed.

When EMD is applied to the synthetic signal, three IMFs and one residue are obtained (**Figure 2**). It is clear that IMF1 does not represent the background cosine wave or the Ricker wavelets singly. Mode mixing occurred seriously in the IMF1. Therefore, the IMF1 lacks physical meaning. Furthermore, affected by the IMF1, the following IMFs are all distorted.

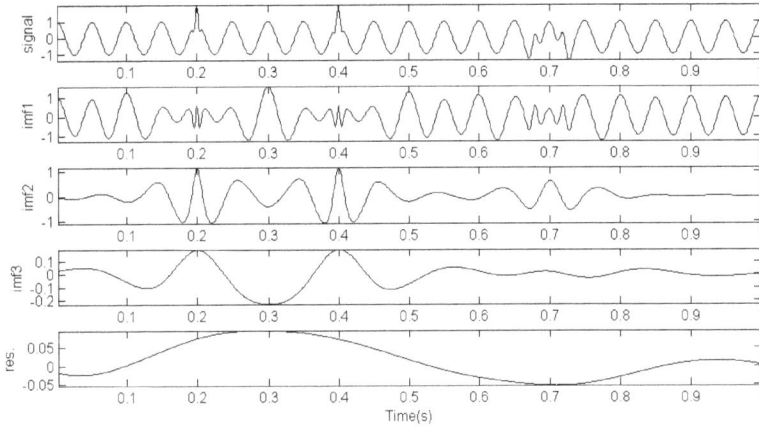

Figure 2. The EMD result of the synthetic signal. Mode mixing is clearly seen in the IMF1. It makes the IMF1 lack the physical meaning. Furthermore, the following IMFs are all distorted.

The SST result is shown in **Figure 3**. We can find that the reconstructed components of the SST better represent the background cosine wave and the Ricker wavelets, respectively. Mode mixing is better suppressed than EMD.

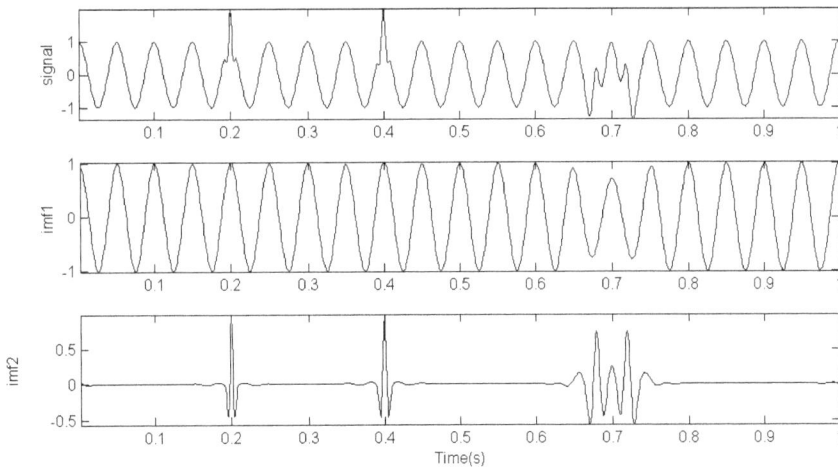

Figure 3. The SST result of the synthetic signal. The reconstructed components of the SST better represent the background cosine wave and the Ricker wavelets, respectively. Mode mixing is better suppressed in SST than in EMD.

Figure 4 shows the VMD result of the synthetic signal. As shown in **Figure 4**, the background cosine wave is first extracted in IMF1. Then the 90 Hz Ricker wavelets at 0.2 s, 75 Hz Ricker wavelets at 0.4 s, and two 40 Hz Ricker wavelets at 0.68 and 0.72 s are mainly retrieved in IMF2. The IMFs of VMD can better represent the individual component of the original signal than the EMD. It shows that the IMFs have more physical meaning.

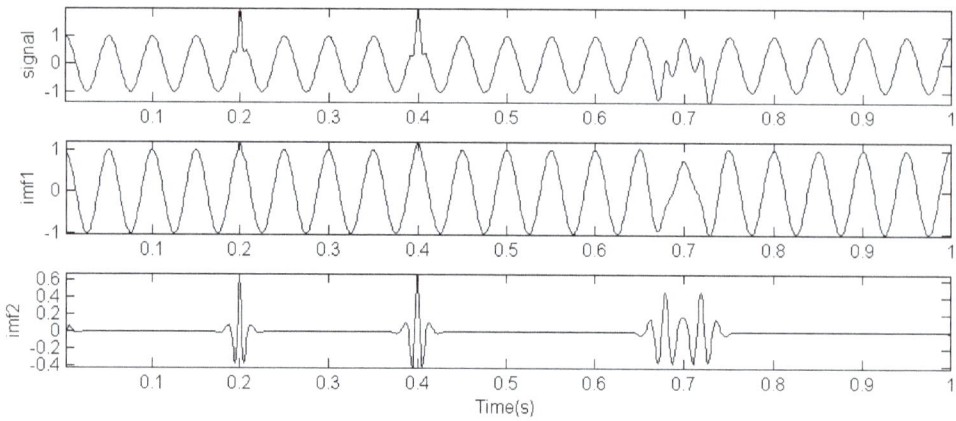

Figure 4. The VMD result of the synthetic signal. The background cosine wave is first extracted in IMF1. Then the Ricker wavelets are mainly retrieved in IMF2. Mode mixing is suppressed the most. The IMFs have more physical meaning.

The spectrums of the EMD output, the SST output, and the VDM output are shown in **Figure 5**. We can find that an overlap of half bandwidth occurred between the two adjacent IMFs for the IMFs in EMD (**Figure 5a**). But for the spectrum of the SST output and the VMD output, the band-pass filters characteristics are shown with the increasing predominant center frequencies (**Figure 5b** and **c**).

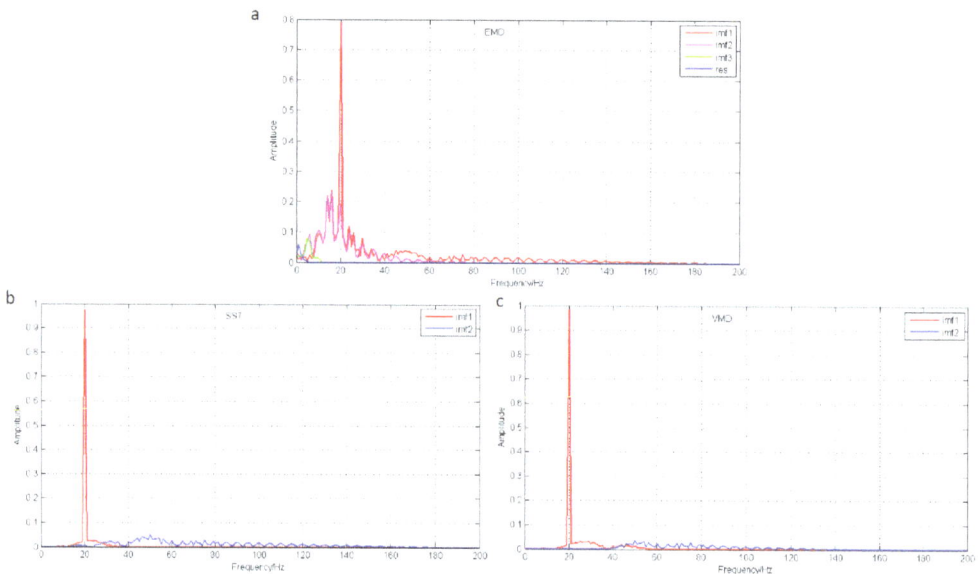

Figure 5. The spectrums of the EMD output (a), the SST output (b), and the VDM output (c). An overlap of half band-width is occurred between the two adjacent IMFs for the IMFs in EMD.

The time-frequency spectrum comparison of the EMD-, SST-, and VMD-based method with short-time Fourier transform (STFT) and wavelet transform is shown in **Figure 6**. Here, the EMD-, SST-, and VMD-based method are respectively using EMD, SST, and VMD combined with Hilbert transform to provide the time-frequency distribution. The EMD-, SST-, and VMD-based method show higher temporal and spatial resolution than STFT and wavelet transform. We also can find that SST- and VMD-based method give more precise time-frequency distribution of the component than EMD-based method.

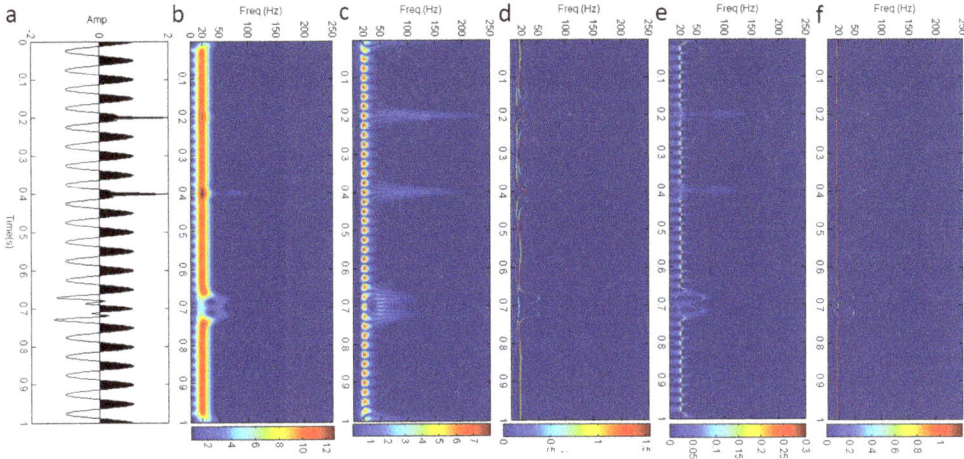

Figure 6. The time-frequency spectrum comparison of the EMD-, SST-, and VMD-based method with short-time Fourier transform (STFT) and wavelet transform for the synthetic data. (a) The synthetic data; (b) STFT; a Hamming window with the length 61 is used. (c) Wavelet transform; continuous wavelet transform with Morlet wavelet is used. (d) EMD-based method; (e) SST-based method; (f) VMD-based method.

To evaluate the noise robustness of the three methods, we use a noisy signal (**Figure 7b**) by adding the Gaussian noise only distributed within the time 0.45–0.55 s (**Figure 7a**) to the synthetic signal in **Figure 1**.

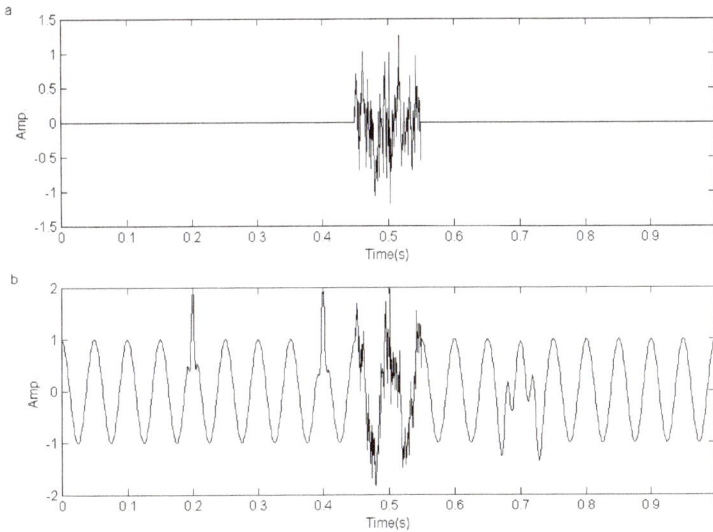

Figure 7. A noisy signal (b) by adding the Gaussian noise only distributed within the time 0.45–0.55 s (a) to the synthetic signal in **Figure 1**.

When EMD is applied to this noisy signal, seven IMFs and one residue are obtained (**Figure 8**). As **Figure 8** shows, the first IMF shows mode mixing phenomenon and it does not represent the either component of the signal. Influenced by the first IMF, the following IMFs are all disturbed and lose their physical meaning.

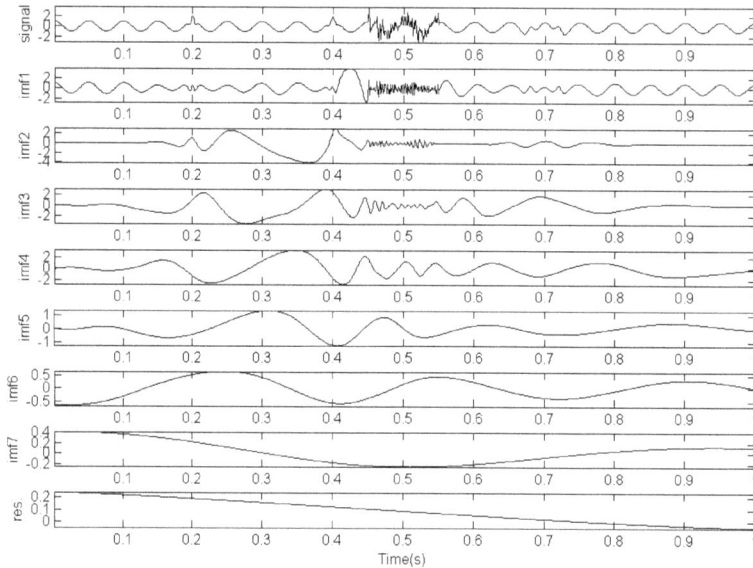

Figure 8. EMD output of the noisy signal. Mode mixing occurred in the first IMF. And influenced by the first IMF, the following IMFs are all disturbed.

Figures 9 and **10** respectively show the SST output and the VMD output of the noisy signal. The background cosine wave is extracted in IMF1 in both the first reconstructed component of SST and the first IMF of the VMD output. The Ricker wavelets are mainly reflected in IMF2 with a slight influence of noise. And the noise is mainly reflected in IMF3. Compared with **Figure 8**, SST and VMD show the noise robustness more than the EMD method.

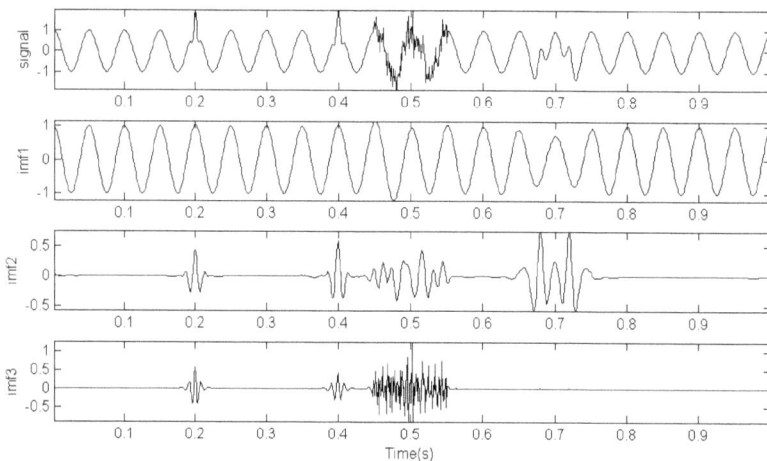

Figure 9. SST output of the noisy signal. The background cosine wave is retrieved in IMF1 in both the first reconstructed component of SST and the first IMF of the VMD output. The noise is mainly reflected in IMF3. The Ricker wavelets are mainly retrieved in IMF2 with a slight influence of noise.

Figure 11 shows the spectrum of the EMD output, the SST output, and the VMD output for the noisy signal. Note that only the spectrum of the first three IMFs are shown in EMD (**Figure 11a**). As shown in **Figure 10(a)**, the bandwidths of IMFs in EMD output are overlapped. Due to the noise influence, the bandwidth of IMF1 becomes larger. But for the spectrums of the SST output and the VMD output (**Figures 11b** and **c**), the bandwidth of IMF1 to IMF2 is similar to that in **Figures 5(b)** and **(c)**. The noise is separated and mainly reflected in IMF3 for the spectrum of the SST output and the VMD output. Compared **Figures 11(b)** and **(c)** with EMD output in **Figure 10(a)**, SST and VMD show better noise robustness.

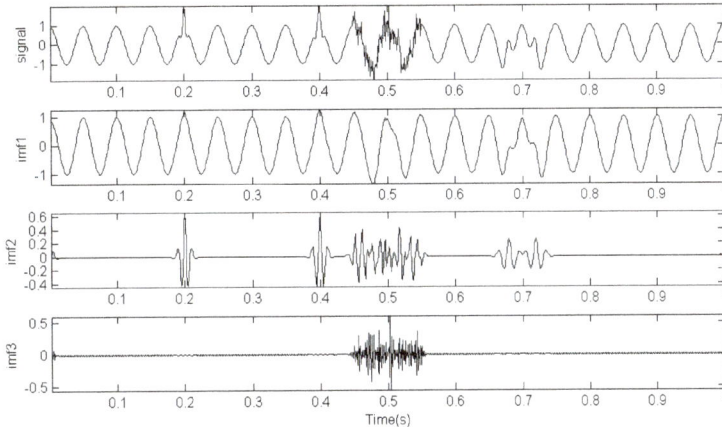

Figure 10. VMD output of the noisy signal. The background cosine wave is mainly extracted in IMF1 in both the first reconstructed component of SST and the first IMF of the VMD output. The noise is mainly reflected in IMF3. The Ricker wavelets are mainly retrieved in IMF2 with a slight influence of noise.

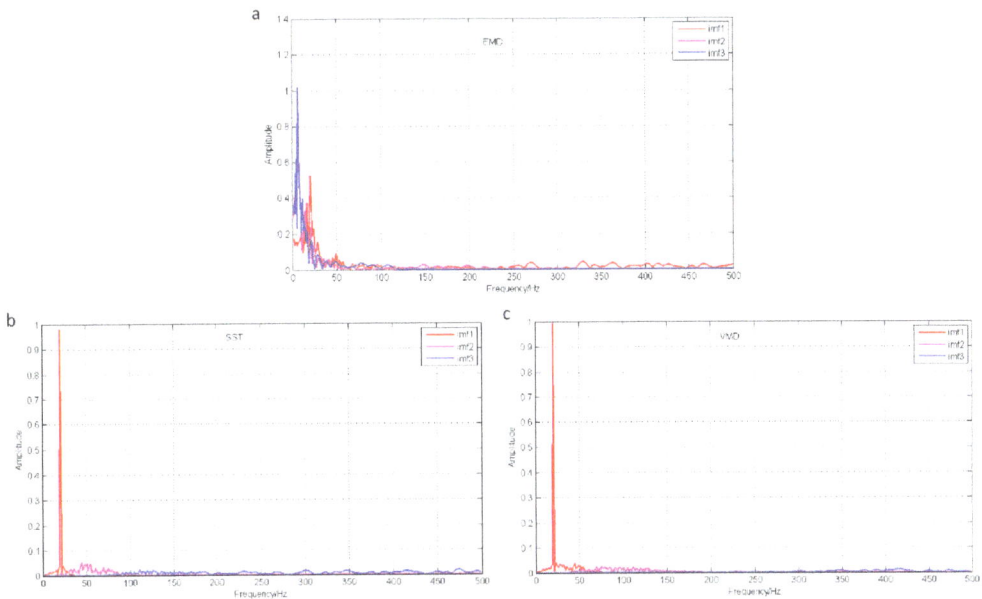

Figure 11. The spectrum of the EMD output (a), the SST output (b), and the VMD (c) output for the noisy signal. SST and VMD show better noise robustness than the EMD does.

3.2. Seismic data

Here, one 2D prestack seismic section with random noise interference from Ordos Basin in China is used for analysis (**Figure 12**). The data is sampled at 1 ms. The SST method is mainly used for seismic data de-noising.

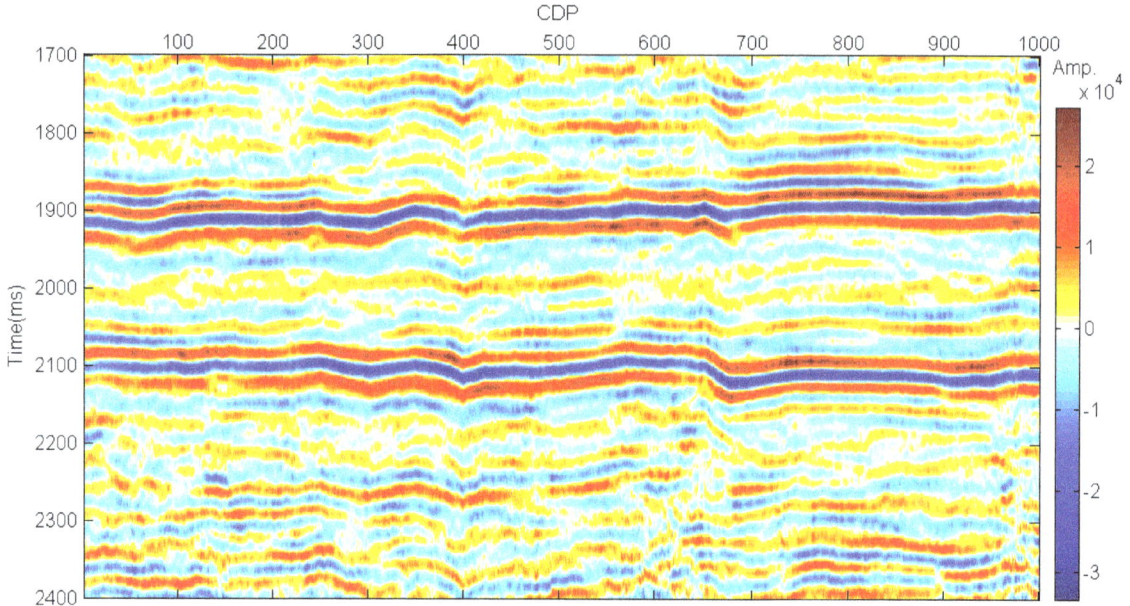

Figure 12. The 2D prestack seismic section with random noise interference from Ordos Basin, China.

By analysis, the main frequency of the seismic section ranges from 0 to 60 Hz. The dominant frequency is 33 Hz. For SST, we use the frequency range [0, 33] to reconstruct the first component and the frequency range [34, 60] for the second component. And the residue frequencies are used to reconstruct the third component. The reconstructed components of the SST for the seismic section are shown in **Figure 13**. Since the random noise always distributes in the high-frequency component, we apply soft thresholding on wavelet transform to the second reconstructed components and abandon the third component which mainly reflects the noise. Then, the de-noising component and the first component are used to generate the de-noising seismic section (**Figure 14**). Compared with the original seismic section in **Figure 12**, the details in the de-noising seismic section are more clear.

3.3. Edge detection of seismic images using LWD

The study on the edge in seismic images has very important significance. Crack, fracture, fault, small rupture, river channel sand boundary, and the boundary of the other special lithology body, and so on all show the edge characteristics in seismic images. In this section, we mainly applied VMD-based method for edge detection in seismic images. The details of 2D-VMD algorithm can be found in reference [26].

Here, the broadband migrated stacked seismic data from a gas field located in western Sichuan Depression, China is collected for analysis (**Figure 15**). If we apply Canny edge detection method to the original seismic section, the output shows more discontinuity and mess (**Figure 16**).

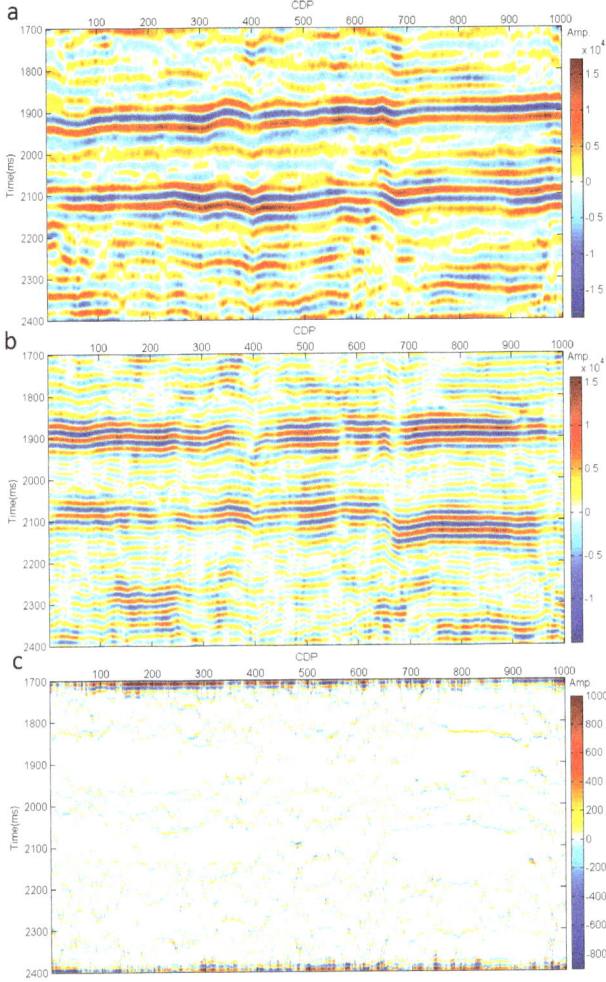

Figure 13. The reconstructed components of the SST for the seismic section. The frequency range [0, 33] is used to reconstruct the first component (a) and the frequency range [34, 60] for the second component (b). The residue frequencies are used to reconstruct the third component (c).

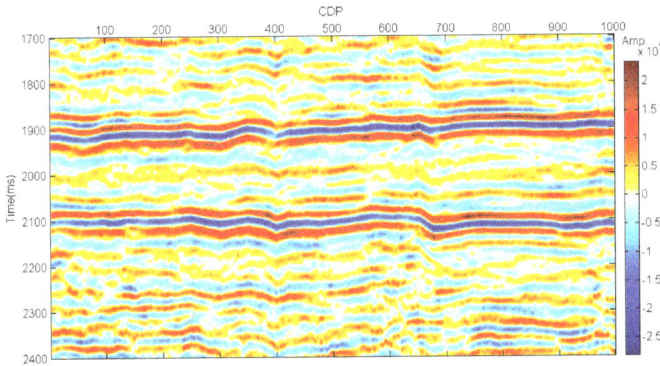

Figure 14. The de-noising seismic section by using SST. Compared with the original seismic section in **Figure 12**, the details in the de-noising seismic section are more clear.

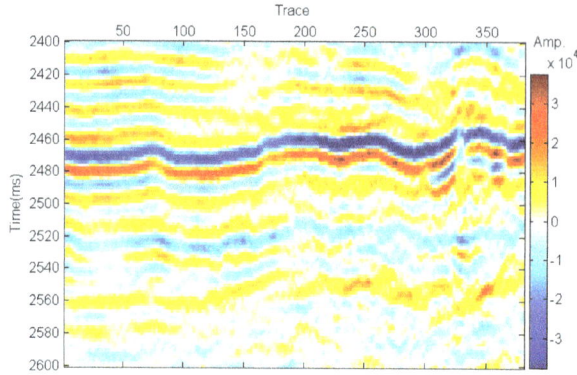

Figure 15. The broadband migrated, stacked seismic section.

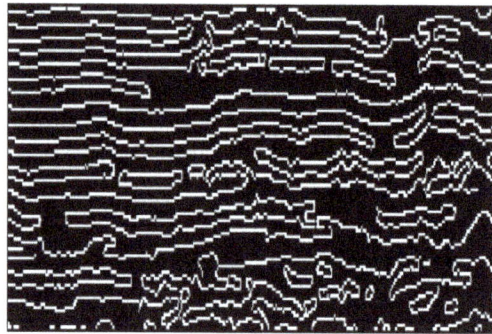

Figure 16. The output of the Canny edge detection for the seismic section.

After 2D-VMD process, the four IMFs are obtained as shown in **Figure 17**. Compare the four IMFs in **Figure 17** with the original seismic section in **Figure 15**, we can find that the second IMF (**Figure 17b**) targets the main reflection information and reflects more details. Then, we use the second IMF for edge detection.

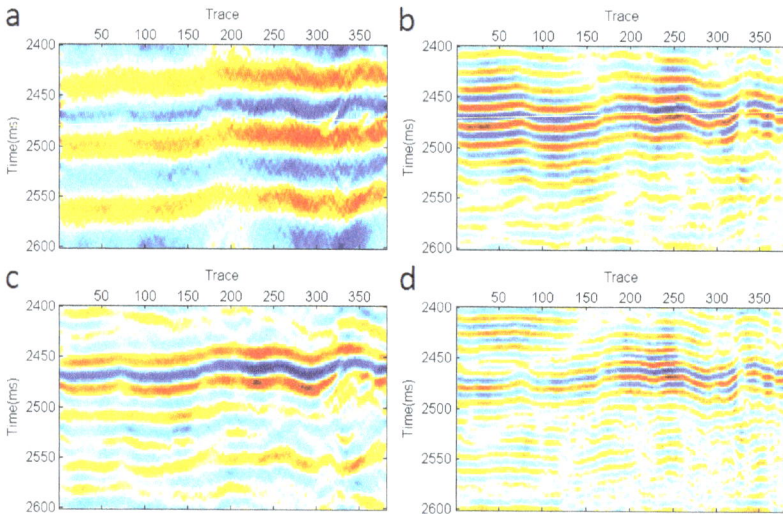

Figure 17. The four IMFs after 2D-VMD for the broadband migrated stacked seismic section. (a) IMF1; (b) IMF2; (c) IMF3; and (d) IMF4.

Next, Canny edge detection is carried out to the second IMF. The output of the Canny edge detection is shown in **Figure 18**. Compared with the results in **Figure 16**, the output of the Canny edge detection applied to the second IMF shows more continuity and the changes in detail and also can be helpful for the automatic seismic event pickup.

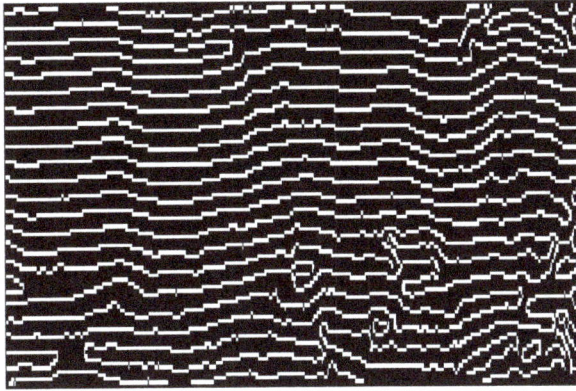

Figure 18. The output of the Canny edge detection for the second IMF.

3.4. Application of LWD on the recovery of the target reflection near coal seams

3.4.1. Model analysis

To evaluate the LWD method for suppressing the strong seismic reflection amplitude of coal seams and recovery of the target reflection near coal seams, one model is designed to simulate the seismic response based on the seismic data and reservoir logging parameters of one tight sandstone reservoir from the Ordos Basin, China.

There are five formations in the geological model 1 (**Figure 19a**). The parameters of each layer are shown in **Table 1**. The layer marked ③ is gas-bearing layer. The layer marked ④ is coal seams. Sampling frequency is 500 Hz. The frequency of the wavelet is 50 Hz. The corresponding seismic response of the geological model is shown in **Figure 19(b)**.

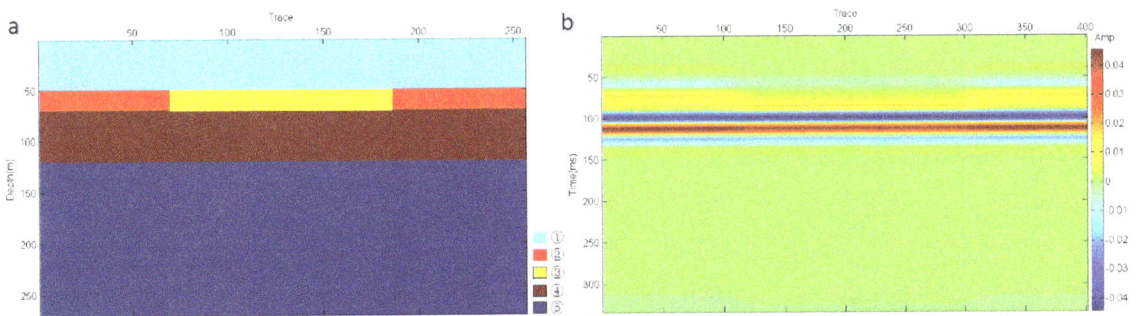

Figure 19. The geological model (a) and the corresponding seismic response (b).

	①	②	③	④	⑤
V (m/s)	3960	4393	4375	4557	3608
Den (kg/m³)	2520	2549	2495	2434	2296

Note: V represents the compressional velocity. Den is the density.

Table 1. The parameters of each layer for the model.

VMD is a data-driven adaptive band-limited filtering method. After VMD process, the original seismic data will be decomposed into a series of IMFs. For the coal-bearing strata coexisted situation, seam stratigraphic features are obvious on one IMF section and weak reflection formation signals are not well represented. Here, the main IMF contributors to coal seams are selected by using the combination of logging information and the maximum correlation. For each selected IMFs, we calculate the energy trace by trace to find out the maximum energy point t_{max}. Then, estimate the main frequency f_d of the IMF of the seismic trace. Time thickness of the thin layer of coal seams is determined by $[t_{max} - k_1 T_d, t_{max} - k_2 T_d]$, where k_1, k_2 are constant factors, respectively, and are estimated by the well-seismic calibration, $T_d = 1/f_d$.

Finally, let $index = \dfrac{E_{ave}}{E_s}$. Using $index$ to suppress the strong amplitude to be consistent with the average energy of the corresponding seismic trace, in which E_s is the energy of the seismic data within the range of $[t_{max} - k_1 T_d, t_{max} - k_2 T_d]$, E_{ave} represents the average energy of the seismic trace.

When all the IMFs that contributed to the main coal seams are processed, sum up the processed IMFs and the remaining IMFs unchanging to reconstruct the seismic trace. Repeat the above process trace by trace, and the strong amplitude suppression section is obtained. This method only suppresses the strong amplitudes of the main coal seams, IMF contributors and the other IMFs, which reflect the remaining information. So the VMD-based method enhances the weak signal components of the coexistence coal-bearing strata.

The result of using the VMD method is shown in **Figure 20**. As shown in **Figure 20**, the strong amplitudes in coal seams are suppressed and the weak reflections near coal seams are enhanced.

3.4.2. Real data

Next, the broadband migrated stacked seismic data from the Ordos Basin, China is collected for analysis, as shown in **Figure 21**. The data is sampled at 2 ms. Strong reflection amplitudes exist in the area where the coal seam is located at around 1900 ms. The sandstone information is with weak amplitudes. The details near coal seams are very weak and evenly hidden by the influence of the strong amplitudes in coal seams.

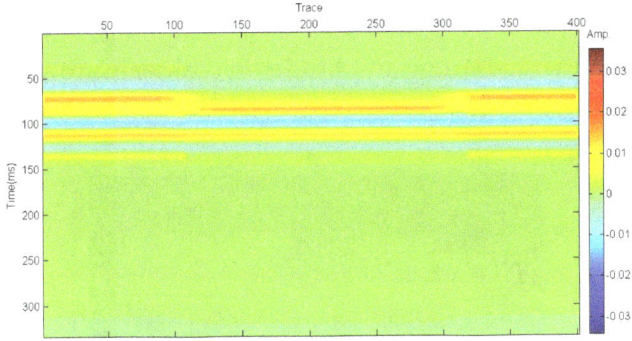

Figure 20. The result of using the VMD method to the model for suppressing the strong amplitudes of the coal seams.

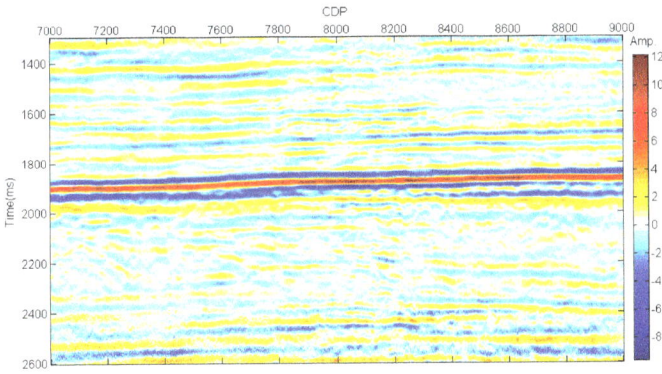

Figure 21. The broadband migrated, stacked seismic data. The data is sampled at 2 ms. There exist strong reflection amplitudes where coal seams are located around 1900 ms. The sandstone information is with weak amplitudes.

The result of using the VMD-based strong amplitude suppression method is shown in **Figure 22**. As shown in **Figure 22**, the strong amplitudes in coal seams are suppressed and the weak reflections near coal seams are enhanced.

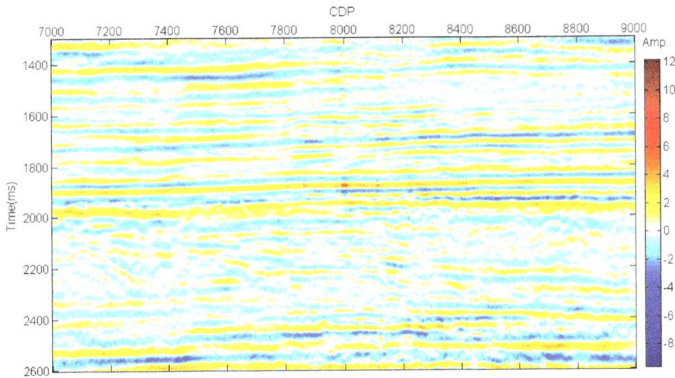

Figure 22. The result of using the VMD method to the seismic section for suppressing the strong amplitudes of the coal seams.

Figure 23 shows the instantaneous amplitudes comparison between the original seismic section and the seismic section after VMD-based strong amplitude suppression method. From **Figure 23(b)**, we can find that the strong amplitudes in coal seam are suppressed most and the details near the coal seams are highlighted.

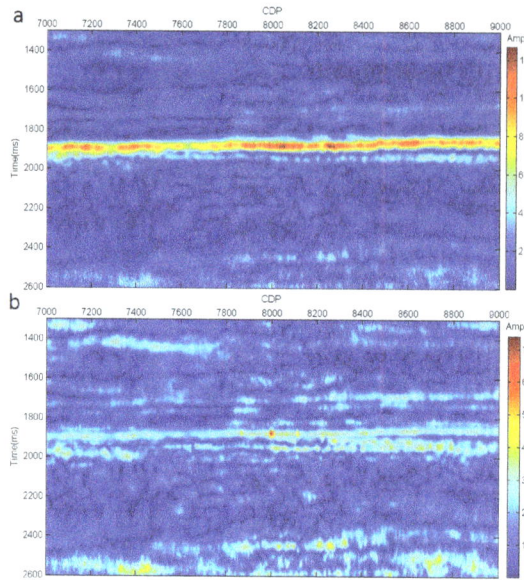

Figure 23. The instantaneous amplitudes comparison between the original seismic section (a) and the seismic section after VMD-based strong amplitude suppression method (b).

Figure 24 shows the instantaneous frequencies comparison between the original seismic section and the seismic section after VMD-based strong amplitude suppression method. We can find that more frequency details are enhanced in the coal seams area around 1900 ms.

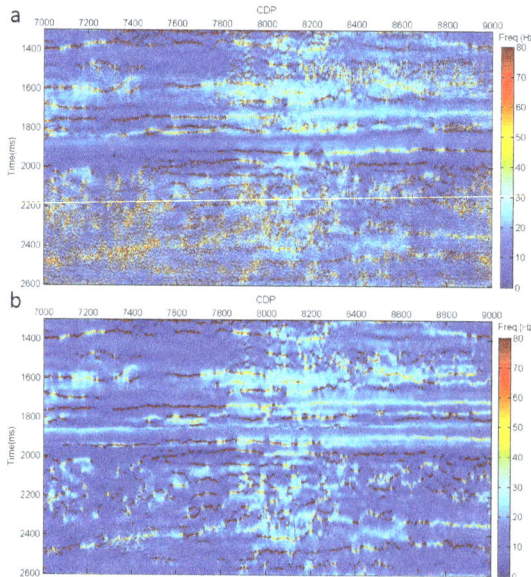

Figure 24. The instantaneous frequencies comparison between the original seismic section (a) and the seismic section after VMD-based strong amplitude suppression method (b).

4. Conclusion

Analysis on synthetic and real data shows that the LWD method is more robust to noise and has stronger local decomposition ability than the traditional Fourier-based and Wavelet-based methods. Comparing with the short-time Fourier transform or wavelet transform, LWD-based time-frequency spectrum promises higher temporal and spatial resolution. Application of the LWD on field data demonstrates that the LWD method can effectively be used in seismic signal de-noising and edge detection. Also, LMD-based method targets the thickness variation of coal seams sensitively and suppresses the strong amplitude in coal seams effectively and highlights the fine details that might escape unnoticed near the area where coal seams are located. The LWD-based technique is more promising for seismic signal processing and interpretation.

Acknowledgements

This work was supported in part by National Natural Science Foundation of China (Grant Nos. 41404102, 41430323, and 41274128) and Sichuan Youth Science and Technology Foundation (Grant No. 2016JQ0012) and in part by Key Project of Sichuan provincial Education Department (No.16ZA0218) and the 2015 annual young Academic Leaders Scientific Research Foundation of CUIT (No. J201507) and the Project of the Scientific Research Foundation of CUIT (No. KYTZ201503). The authors also thank for the supportion of SINOPEC Key Laboratory of Geophysics.

Author details

Ya-juan Xue[1,2*], Jun-xing Cao[2], Gu-lan Zhang[3], Hao-kun Du[4], Zhan Wen[1], Xiao-hui Zeng[1] and Feng Zou[1]

*Address all correspondence to: xueyj0869@163.com

1 School of Communication Engineering, Chengdu University of Information Technology, Chengdu, China

2 School of Geophysics, Chengdu University of Technology, Chengdu, China

3 BGP, CNPC, Zhuozhou, Hebei, China

4 Geophysical Institute of Zhongyuan Oil Field, Sinopec, Henan, China

References

[1] Huang N E, Shen Z, Long S R, et al. The empirical mode decomposition and the Hilbert spectrum for nonlinear and non-stationary time series analysis. Proceedings of the Royal Society of London, Series A: Mathematical, Physical and Engineering Sciences. 1998;454(1971):903–995. DOI: 10.1098/rspa.1998.0193.

[2] Wu Z, Huang N E. Ensemble empirical mode decomposition: A noise-assisted data analysis method. Advances in Adaptive Data Analysis. 2009;1(01):1–41. DOI: 10.1142/S1793536909000047.

[3] Torres M E, Colominas M A, Schlotthauer G, et al. A complete ensemble empirical mode decomposition with adaptive noise. In: 2011 IEEE International Conference on Acoustics, Speech and Signal Processing (ICASSP); 22–27 May 2011; Prague, Czech Republic. IEEE; 2011. pp. 4144–4147.

[4] Daubechies I, Lu J, Wu H T. Synchrosqueezed wavelet transforms: An empirical mode decomposition-like tool. Applied and Computational Harmonic Analysis. 2011;30(2): 243–261. DOI: 10.1016/j.acha.2010.08.002.

[5] Dragomiretskiy K, Zosso D. Variational mode decomposition. IEEE Transactions on Signal Processing. 2014;62(3):531–544. DOI: 10.1109/TSP.2013.2288675.

[6] Barnhart B L, Eichinger W E. Empirical mode decomposition applied to solar irradiance, global temperature, sunspot number, and CO_2 concentration data. Journal of Atmospheric and Solar-Terrestrial Physics. 2011;73(13):1771–1779. DOI: 10.1016/j.jastp.2011.04.012.

[7] Xue Y, Cao J, Tian R. A comparative study on hydrocarbon detection using three EMD-based time-frequency analysis methods. Journal of Applied Geophysics. 2013;89:108–115. DOI: 10.1016/j.jappgeo.2012.11.015.

[8] Liu B, Riemenschneider S, Xu Y. Gearbox fault diagnosis using empirical mode decomposition and Hilbert spectrum. Mechanical Systems and Signal Processing. 2006;20(3):718–734. DOI: 10.1016/j.ymssp.2005.02.003.

[9] Chatlani N, Soraghan J J. EMD-based filtering (EMDF) of low-frequency noise for speech enhancement. IEEE Transactions on Audio, Speech, and Language Processing. 2012;20(4):1158–1166. DOI: 10.1109/TASL.2011.2172428.

[10] Andrade A O, Nasuto S, Kyberd P, et al. EMG signal filtering based on empirical mode decomposition. Biomedical Signal Processing and Control. 2006;1(1):44–55. DOI: 10.1016/j.bspc.2006.03.003.

[11] Mostafanezhad I, Boric-Lubecke O, Lubecke V, et al. Application of empirical mode decomposition in removing fidgeting interference in Doppler radar life signs monitoring devices. In: 2009 Annual International Conference of the IEEE Engineering in

Medicine and Biology Society; 3–6 Sept. 2009; Minnesota, USA. IEEE; 2009. pp. 340–343.

[12] Lin D C, Guo Z L, An F P, et al. Elimination of end effects in empirical mode decomposition by mirror image coupled with support vector regression. Mechanical Systems and Signal Processing. 2012;31:13–28. DOI: 10.1016/j.ymssp.2012.02.012.

[13] Rilling G, Flandrin P, Goncalves P. On empirical mode decomposition and its algorithms. In: The 2003 IEEE-EURASIP workshop on nonlinear signal and image processing; Trieste, Italy. IEEER; 2003. pp. 8–11.

[14] Flandrin P, Rilling G, Goncalves P. Empirical mode decomposition as a filter bank. IEEE Signal Processing Letters. 2004;11(2):112–114. DOI: 10.1109/LSP.2003.821662.

[15] Rilling G, Flandrin P. On the influence of sampling on the empirical mode decomposition. In: 2006 IEEE International Conference on Acoustics, Speech and Signal Processing (ICASSP) (3); 14–19 May 2006; Toulouse, France. IEEE; 2006. pp. 444–447.

[16] Huang N E, Wu Z. A review on Hilbert-Huang transform: Method and its applications to geophysical studies. Reviews of Geophysics. 2008;46(2):RG2006. DOI: 10.1029/2007 RG000228.

[17] Mandic D P, ur Rehman N, Wu Z, et al. Empirical mode decomposition-based time-frequency analysis of multivariate signals: The power of adaptive data analysis. IEEE Signal Processing Magazine. 2013;30(6):74–86. DOI: 10.1109/MSP.2013.2267931.

[18] Li C, Liang M. Time-frequency signal analysis for gearbox fault diagnosis using a generalized synchrosqueezing transform. Mechanical Systems and Signal Processing. 2012;26:205–217. DOI: 10.1016/j.ymssp.2011.07.001.

[19] Huang N E, Wu Z, Long S R, et al. On instantaneous frequency. Advances in Adaptive Data Analysis. 2009;1:177–229. DOI: 10.1142/S1793536909000096.

[20] Xue Y-J, Cao J-X, Tian R-F. EMD and Teager-Kaiser energy applied to hydrocarbon detection in a carbonate reservoir. Geophysical Journal International. 2014;197:277–291. DOI: 10.1093/gji/ggt530.

[21] Wang T, Zhang M, Yu Q, et al. Comparing the applications of EMD and EEMD on time-frequency analysis of seismic signal. Journal of Applied Geophysics. 2012;83:29–34. DOI: 10.1016/j.jappgeo.2012.05.002

[22] Daubechies I. Ten lectures on wavelets. In: CBMS-NSF Regional Conference Series in Applied Mathematics; 1992. Philadelphia: Society for industrial and applied mathematics. Vol.61, pp.198–202.

[23] Li C, Liang M. A generalized synchrosqueezing transform for enhancing signal time-frequency representation. Signal Processing. 2012;92:2264–2274. DOI: 10.1016/j.sigpro. 2012.02.019.

[24] Wu H-T, Flandrin P, Daubechies I. One or two frequencies? The synchrosqueezing answers. Advances in Adaptive Data Analysis. 2011;3(2):29–39. DOI: 10.1142/S179353691100074X.

[25] Xue Y-J, Cao J-X, Wang D-X, et al. Application of the variational mode decomposition for seismic time-frequency analysis. IEEE Journal of Selected Topics in Applied Earth Observations and Remote Sensing. Forthcoming. 2016;9(8):3821-3831.

[26] Dragomiretskiy K, Zosso D. Two-dimensional variational mode decomposition. In: International Workshop on Energy Minimization Methods in Computer Vision and Pattern Recognition. 13-16 January 2015; Hong Kong, China. pp. 197–208.

6

Earthquake Instrumentation

Jae Cheon Jung

Additional information is available at the end of the chapter

Abstract

Earthquake detectors (seismic instruments) are used to measure low-frequency ground motion caused by earthquakes. They detect the seismic waves created by ground motions and convert the wave motions into electronic signals, which are measurable. Two measurement principles are widely used in the industry sector to detect the strong motion of earthquake that require high sensitivity: force-balanced acceleration and servo acceleration with a feedback loop. In this chapter, the motion of the mass as a function of the ground displacement is discussed with a differential equation resulting from the equilibrium of forces. In addition, the transfer functions of both instruments are investigated by using Matlab® Simulink. This technology is applied in NPP (nuclear power plant) to ensure the safety of the plant in systems, such as the SMS (seismic monitoring system) and the ASTS (automatic seismic trip system). SMS provides monitoring and recording capability, whereas ASTS makes a decision to trip the NPP when the PGA (peak ground acceleration) exceeds the pre-defined value, which is determined based on the ground conditions.

Keywords: earthquake detector, seismic instruments, accelerometer, force balanced acceleration, servo acceleration, transfer function

1. Introduction

Earthquake detection instrumentation is widely used in many industries to secure the system, structure and components. In addition, early warning of incident seismic wave is critical to safety of human lives. On 12 September 2016, South Korea experienced the most powerful earthquake ever recorded in the country since measurements began in 1978. A 5.8-magnitude earthquake struck the historic city of Gyeongju and the people were subjected to a series of aftershocks affecting their daily lives [1]. Buildings built inside the historic district were not strong enough to sustain seismic incident despite its low destructive power. Luckily, since the frequency of the wave was higher than 16 Hz, a catastrophic falling down of buildings was not experienced. This tells the relationship among magnitude, wave frequency and the destructive power. Earthquake detection instrumentation has been improved to detect the magnitude and frequency effectively. Historically, the first seismic instrument was invented in Han dynasty. Modern instruments are based on electronic sensors which are able to detect tri-axial acceleration.

In order to investigate both theory and application of earthquake instruments, the mathematical expression of the instrument is defined by the motion of the mass-spring-damper system. Then the differential equation resulting from the equilibrium of forces is discussed. In addition, the transfer functions of both force balanced and servo balanced and feedback circuit are analysed. Finally, the application of earthquake instrument in the nuclear power plant is introduced.

2. Mathematical expression

Let us define a single seismic instrument having mass, spring and dashpot as shown in **Figure 1**. A dashpot is a device that provides viscous friction or damping [2]. The dashpot absorbs energy. This spring-mass-dashpot system can be expressed using Newton's second law as represented in Eq. (1):

$$\text{(D.E.)} \; m\ddot{y} = -ky - b\dot{y} + u \tag{1}$$

$$\text{(T.F.)} \; ms^2 Y(s) = -kY(s) - bsY(s) + U(s) \tag{2}$$

$$G(s)\frac{Y(s)}{U(s)} = \frac{1}{ms^2 + bs + k} \tag{3}$$

The mathematical expression of the system above can be expressed as Eq. (1). The transfer function of Eq. (1) can be obtained as Eqs. (2) and (3) by taking Laplace transformation.

The motion of the mass as a function of the ground displacement is given by a differential equation resulting from the equilibrium of forces as shown in Eq. (4) [3].

$$Fs + Fr + Fg = 0 \tag{4}$$

where $Fs = -kx$ (k is the spring constant);

$Fr = -bx$ (b is the friction coefficient);

$Fg = -mu$ (m is the spring mass).

When the seismic instrument moves slowly, the acceleration and velocity become negligible, it records ground acceleration. If however, movement is fast enough, the acceleration of the mass dominates and the ground displacement is recorded.

Figure 2 shows the seismic instrument under ground motion. Where the u_g is the ground displacement, x_r is displacement of the seismometer mass and x_0 is the mass equilibrium position.

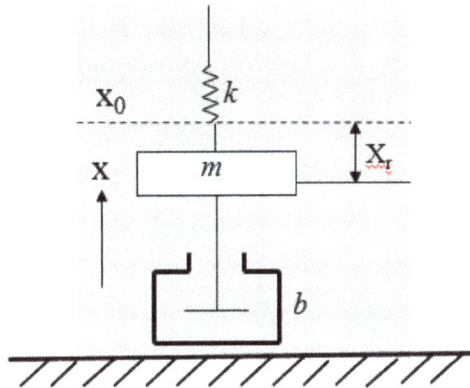

Figure 1. Spring-mass-dashpot system.

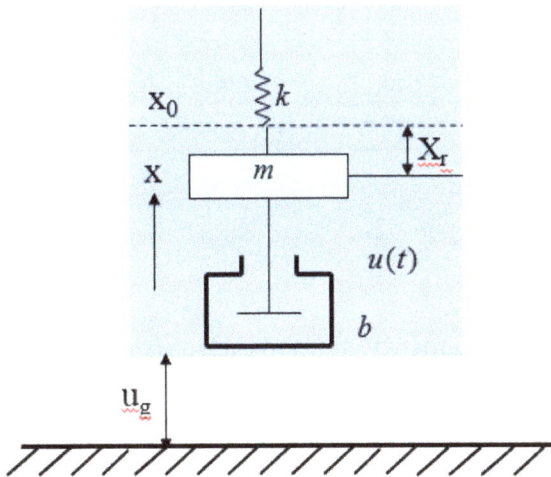

Figure 2. Concept of seismic instrument under ground motion.

In general, the frequency of the seismic wave tends to be increased if the magnitude of the wave is getting smaller. The ground motion has very low-frequency profile due to the movement of the earth.

This study is focused on the plant industry application of the earthquake instrument. In this case, the detector covers DC to 50 Hz frequency components.

3. Tri-axial accelerometer

The tri-axial accelerometer has been widely used in industrial applications. An accelerometer is a sensor which can detect the earthquake-induced motion. The tri-axial accelerometer has a coil in the sensor that is suspended to it. When the accelerometer detects the earthquake-induced motion, the coil is moved and the voltage is induced proportionally to the variation of magnetic force. This voltage is converted to acceleration.

The tri-axial accelerometer detects incident seismic waves from longitudinal (L), transverse (T) and vertical (V) direction using three identical sensors as illustrated in **Figure 3**. In general, each accelerometer has a feedback amplifier in consideration of the stability of the sensor output. Moreover, an added RC filter eliminates the high-frequency component from the output. The accelerometers are installed at free-field where the earthquake ground motion is detected by the surface motion.

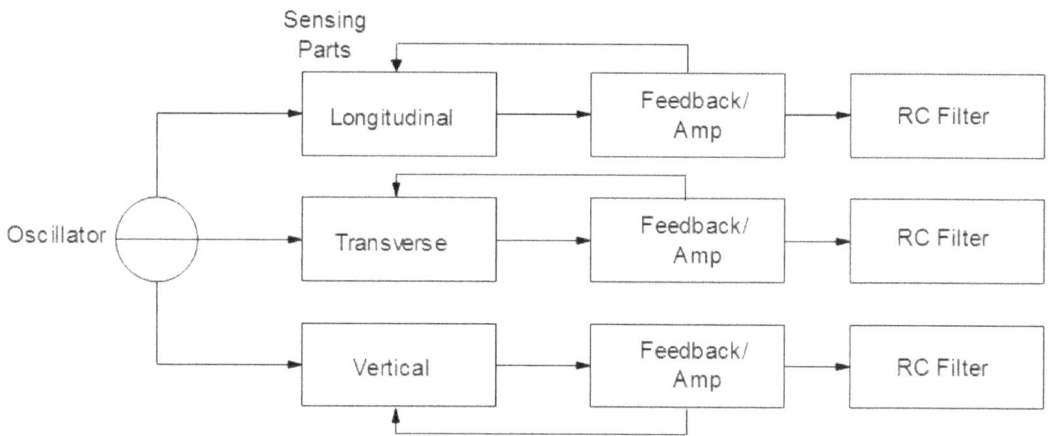

Figure 3. Concept of tri-axial accelerometer.

3.1. Types of accelerometers

An accelerometer detects low-frequency component from the earthquake ground motion. Two measurement principles are used to detect a strong motion earthquake requiring high sensitivity: open-loop type or closed-loop type with a feedback loop. In the plant industry, two sensor types have been used; force balanced acceleration using a pendulum, and servo acceleration with a feedback loop.

3.1.1. Force balanced acceleration using pendulum

In this section, the FBA-23 (by Kinemetrics Inc.) accelerometer is analysed to provide a more detailed operational theory of pendulous type servo sensors. As seen in **Figure 4**, FBA-23 consists of three-axis FBAs (force balanced accelerators) for detecting incident seismic waves in (L), (T) and (V) directions. Each accelerometer is identical, however, the vertical one has an active oscillator circuit as an additional function.

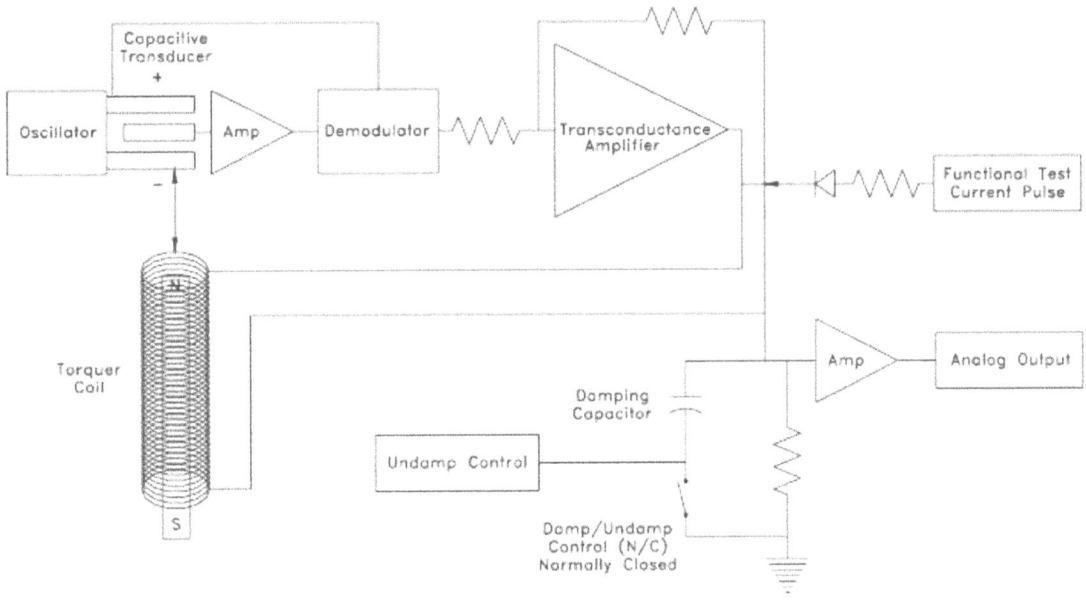

Figure 4. Simplified circuit of FBA-23 [4].

The electrical circuit of FBA-23 has a non-inverting amplifying circuit for adjusting the polarity of the output signal. The transfer function of the accelerometer can be expressed by Eq. (5).

$$\frac{E_{OUT}}{E_{DC}} = \frac{\omega_n^2}{s^2 + 2\zeta\omega_n s + \omega_n^2} \tag{5}$$

where ω_n = natural frequency; s = Laplacian; ζ = damping factor.

The solution of Eq. (5) can be denoted by the second order equation having two poles $(p_1, \ p_2)$ as shown in Eqs. (6) and (7).

$$p_{1=} - \zeta\omega_n + j\omega_n\sqrt{1-\zeta} \tag{6}$$

$$p_{2=} - \zeta\omega_n - j\omega_n\sqrt{1-\zeta} \tag{7}$$

The circuit has sensitivities of 0.25, 0.5, 1, 2 and 4 g full scale. This corresponds to normal absolute sensitivity at DC.

The second order transfer function in Eq. (5) has natural frequencies of 50, 90 or 100 Hz. In addition, if we consider the damping factor of 0.707, then the two poles (p_1, p_2) are located in the left half of the s-plane. The transfer function of this accelerometer can be rewritten as Eq. (8) having two poles.

$$\frac{E_{OUT}}{E_{DC}} = \frac{\omega_n^2}{s^2 + 2\zeta\omega_n s + \omega_n^2} = \frac{\omega_n^2}{(s - P1)(s - P2)} \tag{8}$$

From Eq. (8), the natural frequency and damping factor of this instrument can be calculated as Eqs. (9) and (10) respectively.

$$\omega_n = 2 \cdot \pi \cdot 50 = 314.159 \tag{9}$$

$$\zeta = 0.2550 (\text{for } 1.0 \text{ g}) \tag{10}$$

The poles are then calculated as, $P1$ = -222.1 + j222.1 rad/s and $P2$ = -222.1 - j222.1 rad/s. Therefore, the resultant transfer function of this system becomes Eq. (11)

$$TF = \frac{98696}{s^2 + 160.22 s + 98696} \tag{11}$$

Figure 5 shows the Matlab Simulink results when the step input is applied. The corresponding plot described in **Figure 6** shows overshoots and it needs relatively reduced settling time. In this simulation, the poles have 50 Hz of natural frequency and 1 g of absolute gain.

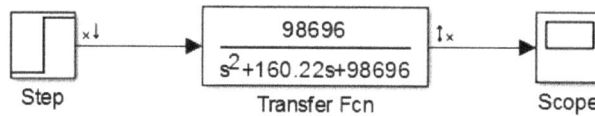

Figure 5. Matlab Simulink diagram (without RC filter).

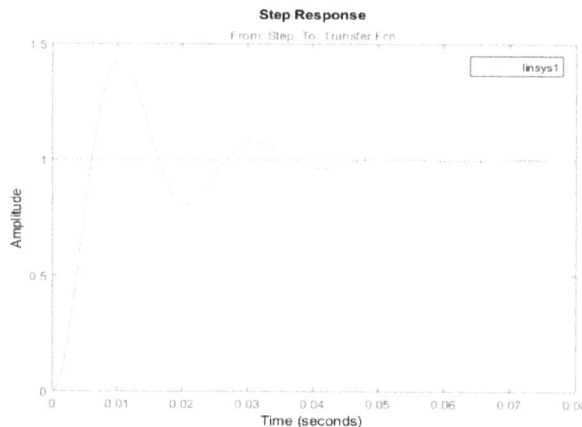

Figure 6. Step response (without RC filter).

Since this transfer function may be affected by the zeros, therefore by adding low pass RC filter to the post amplifier, the stability and accuracy of this circuit would be improved. In this case, the RC filter adds one more pole as summarized in **Table 1**. This proves that the calculated poles are placed in the left half of the *s*-plane as well.

	P1	P2	P3
Real (rad/s)	−222.1	−222.1	−1000
Imaginary (rad/s)	222.1	−222.1	0

Table 1. Pole placement of FBA-23 in 50 Hz and 1 g

The full transfer function of the electrical circuit in 50 Hz has three poles as shown in Eq. (12).

$$TF_{RC} = \frac{98696}{s^3 + 1160.225\,s^2 + 258921s + 98696000} \tag{12}$$

Figure 7 shows the Matlab Simulink results when a step input is applied. **Figure 8** also shows almost the same plot disregarding the existence of the RC filter.

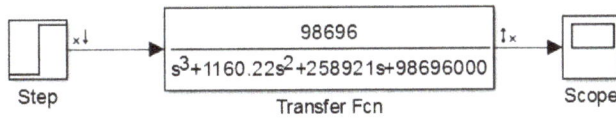

Figure 7. Matlab Simulink diagram (with RC filter).

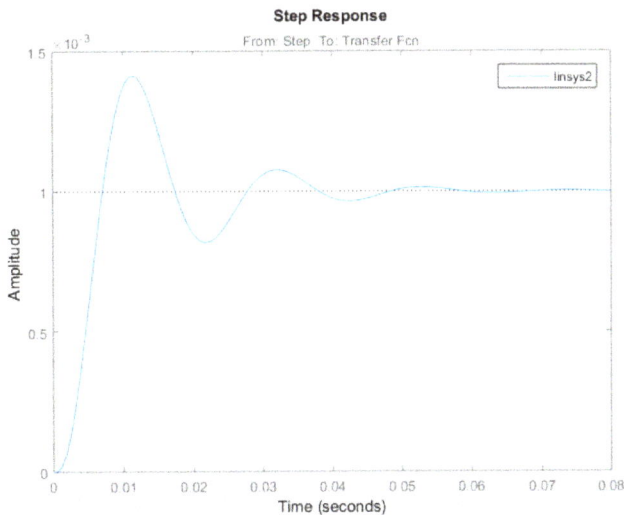

Figure 8. Step response (with RC filter).

3.1.2. Servo acceleration with feedback loop

In this section, the servo acceleration with a feedback loop type accelerometer (AC-23 by Geosig), which consists of geophone and electrical circuits, is discussed. The geophone response can be seen from two viewpoints:

- With a constant velocity, amplitude applied to the geophone.

- With a constant acceleration, amplitude applied to the geophone.

Figure 9 represents the response curve of Geophone at 4.5 Hz.

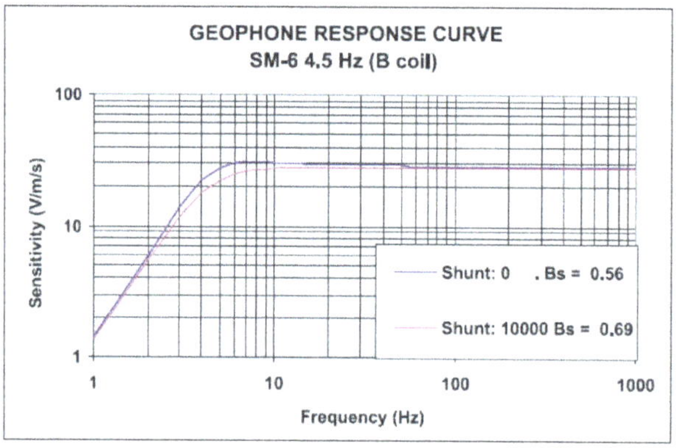

Figure 9. Geophone response curve (4.5Hz) [5].

The relationship between these two points of view is a rotation when plotted with log-log axis scales. In **Figure 10**, the principle of Geophone is drawn. When the motion on the *x*-axis is incident then Geophone generates the open-loop signal which has peak response of 4.5 Hz.

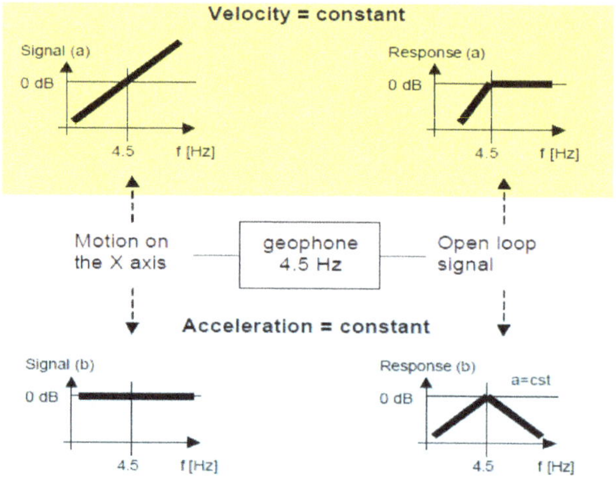

Figure 10. Principle of Geophone [6].

The geophone is a long coil travel version and the extra coil travel offers an advantage for higher tilt requirements where larger amplitude signals may be encountered. A range of natural frequencies is available from 4.5 Hz to 14 Hz, providing a choice of the correct geophone for a wide variety of applications. The geophone is suitable for detecting an extensive range of ground motion.

The AC-23 provides over-damping the geophones with a feedback amplifier in a bridge circuit. The principle of over-damping geophones is done by applying a voltage on the geophone, which has opposite polarity from the voltage, which is induced by the moving geophone coil. Since the voltage induced by the geophone coil is proportional to the velocity, the externally applied voltage is also proportional to it. This results in a current in the coil (~force), which is also proportional to the velocity and therefore is a "damping" current, or additional damping. Increasing this damping further will lead to a resulting output that is proportional to acceleration.

The function of the sensor, seen from a constant acceleration point of view is illustrated in **Figure 11**.

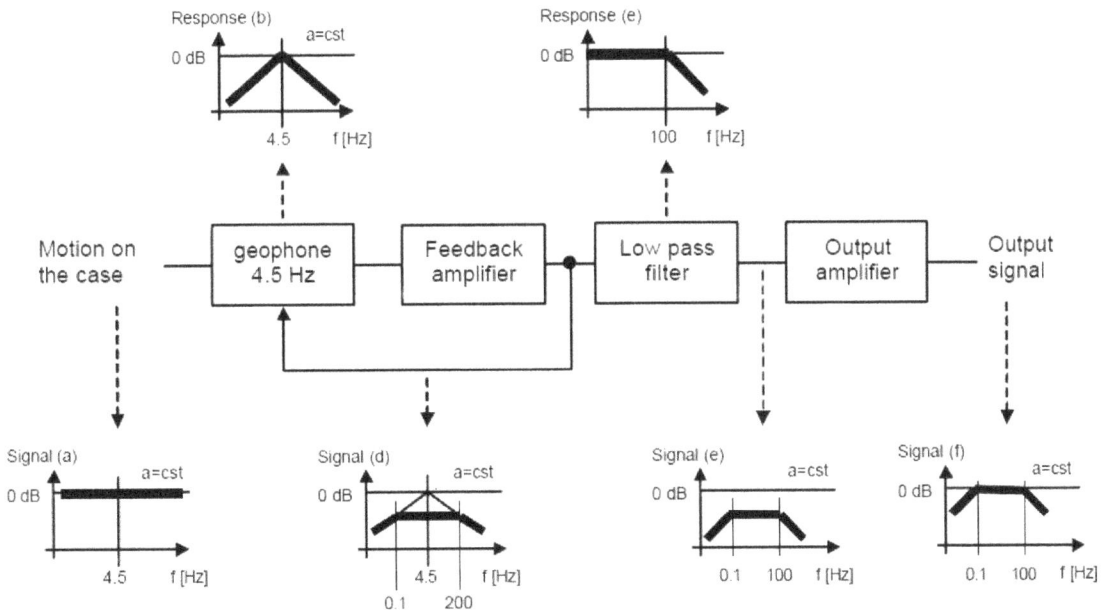

Figure 11. Operational principle of AC-23 [6].

The geophone is connected in a resistor bridge, driven by a feedback amplifier and an inverter, which apply the amplified bridge differential signal in opposite polarity [6]. The amplifier has a fixed gain that will define the bandwidth of the accelerometer. The gain G and output V_o are represented by Eqs. (13) and (14). **Figure 12** shows the equivalent circuit for the linear servo-balanced accelerometer.

$$G = 1 + \left(\frac{Z_2}{Z_1} \right) \tag{13}$$

$$V_0 = G \cdot V_i = \left(1 + \frac{Z_2}{Z_1} \right) \cdot V_i \tag{14}$$

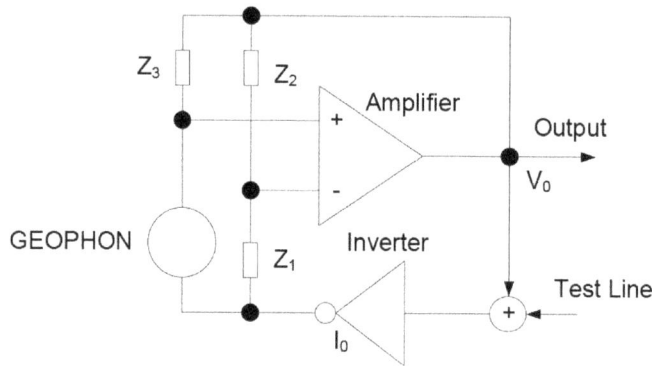

Figure 12. Equivalent circuit for AC-23 (linear servo balanced accelerometer) [6].

In **Figure 12**, the test-line shifts the voltage to one side of the bridge, which produces a current flow in the geophone, resulting in a displacement step of the seismic mass. The movement of the mass generates a voltage across the Geophone, which is detected by the differential amplifier and induces an output signal. The effect of the test signal on the bridge is cancelled by the differential input of the amplifier.

In this part, the Geophone transfer function transfer is analysed in order to define how the sensor generates open-loop transfer function as depicted in Eq. (13).

$$TF = \frac{1.564s}{0.1251s^2 + 1.514s + 1} \tag{15}$$

The transfer function in Eq. (15) has two poles and one zero. In this work, the MATLAB script of Geophone is used to illustrate the response curve under step input condition as shown in **Figure 13**.

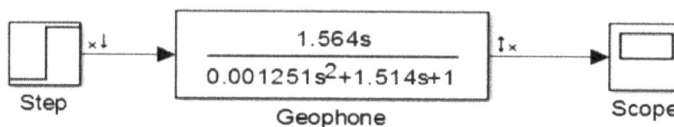

Figure 13. Open-loop transfer function of Geophone.

The transfer function plotted as the frequency range vs. relative response to the acceleration is shown in **Figure 14**.

If a unit step function is applied to the transfer function, the response as in **Figure 15** is obtained.

In this case, the magnitude and the phase can be drawn by the Bode plot as seen in **Figure 16**.

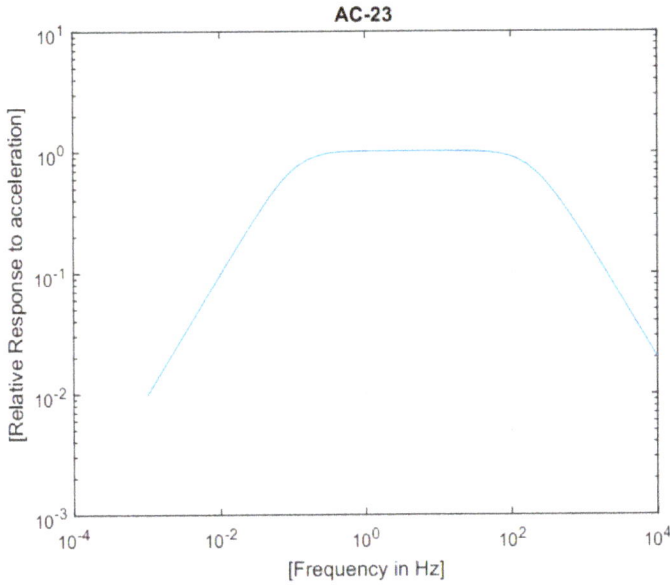

Figure 14. Transfer function plotted by the frequency vs. relative response to the acceleration.

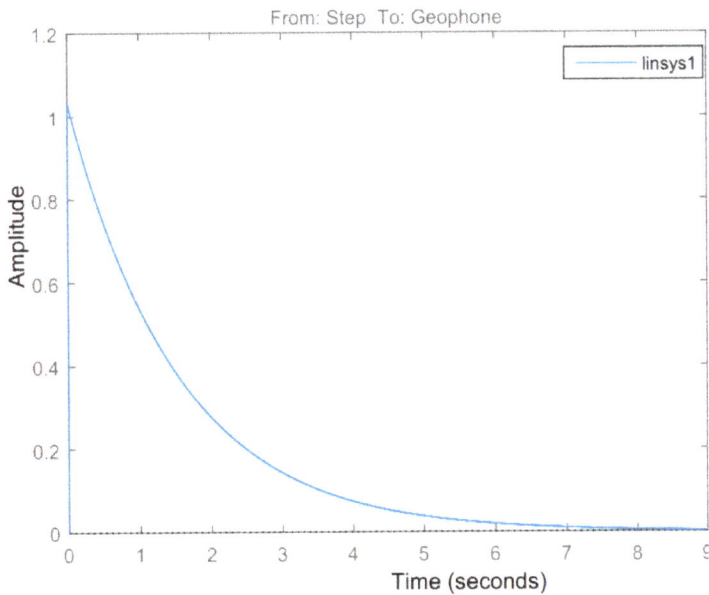

Figure 15. Step response.

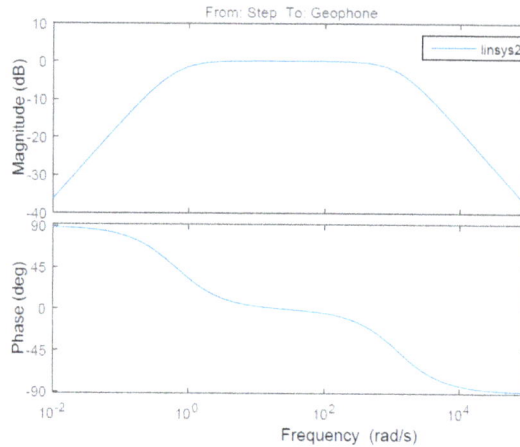

Figure 16. Bode plot.

4. Application of seismic instrument

In this chapter, two application examples for the NPP (nuclear power plant) are discussed with their designs. They are SMS (seismic monitoring system) and ASTS (automatic seismic trip system). SMS detects the seismic wave and provides alarms to the NPP operators. The operator then makes a decision whether or not to trip the nuclear reactor. ASTS has an automatic trip function, if the peak acceleration of the incident wave exceeds, the pre-determined level trigger the reactor trip automatically. In the protection view-point, SMS provides alarms for the manual trip while ASTS provides automatic protection.

4.1. Seismic monitoring system

The SMS consists of tri-axial accelerometers and a seismic monitoring cabinet which is located in an electrical equipment room in unit one (only one unit). The cabinet contains recorders, a playback and analysis unit, an annunciator, a seismic switch, test units and an uninterruptible power supply.

The signal flow from the accelerometer to the seismic monitoring system is shown in **Figure 17**. From the accelerometer, the voltage signal is transmitted to the seismic monitoring cabinet and is converted by electronics into a signal proportional to the acceleration. The acceleration signal can activate recorders and/or local and MCR annunciators/alarms according to its magnitude.

When the acceleration of earthquake exceeds the seismic trigger or set-points for OBE (Operating Basis Earthquake or Event) or SSE (Safe Shutdown Earthquake), the alarm is provided on the local annunciator panel of the seismic monitoring cabinet and that of large panel display system (LDPS) in main control room or alternatively in the remote shutdown room. That event also activates the time-history recorder(s) in the seismic monitoring cabinet.

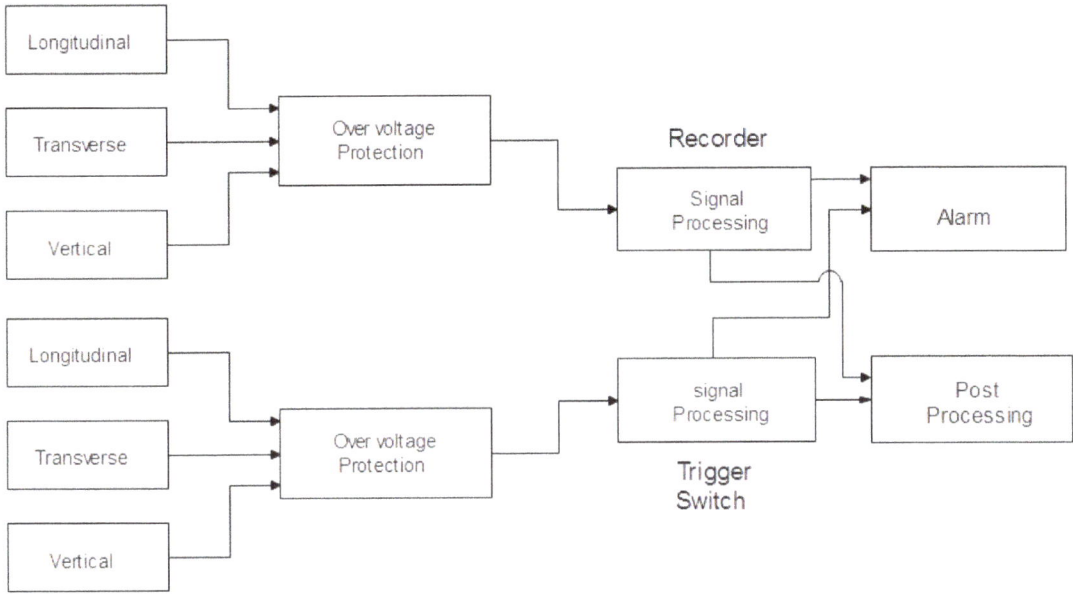

Figure 17. SMS block diagram.

4.2. Automatic seismic trip system

ASTS continuously monitor PGA (peak ground acceleration) from the seismic wave and automatically generates the trip signal. As shown in **Figure 18**, the seismic signal detected by the accelerometer is filtered, rectified and converted to the current. The sensor module is to cut off the frequency ranges over 10 Hz in order to pick out the strong motion of the earthquake. Since the seismic signal has a positive and negative signal, a rectifier circuit is added. The rectifier converts the bipolar signal to the unipolar.

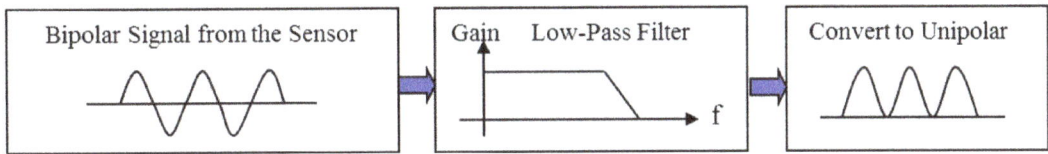

Figure 18. Conceptual diagram of sensor module for signal processing [7].

The overall scheme of ASTS is shown in **Figure 19**. The sensor output is interfaced with digital input card at the ASTS cabinet. Even the decision logic is set as two out of four but the system is composed of two independent channels. They are N1 and N2. For isolation between the channels, the digital input/output card is applied. The dotted line on the ASTS cabinet is the scope of trip logic channel N1 and N2. The ASTS cabinet is implemented by digitalized system such as PLC (programmable logic controller), FPGA (field programmable gate array), or DCS (distribute control system). The bi-stable and decision logic are configured with software inside each channel.

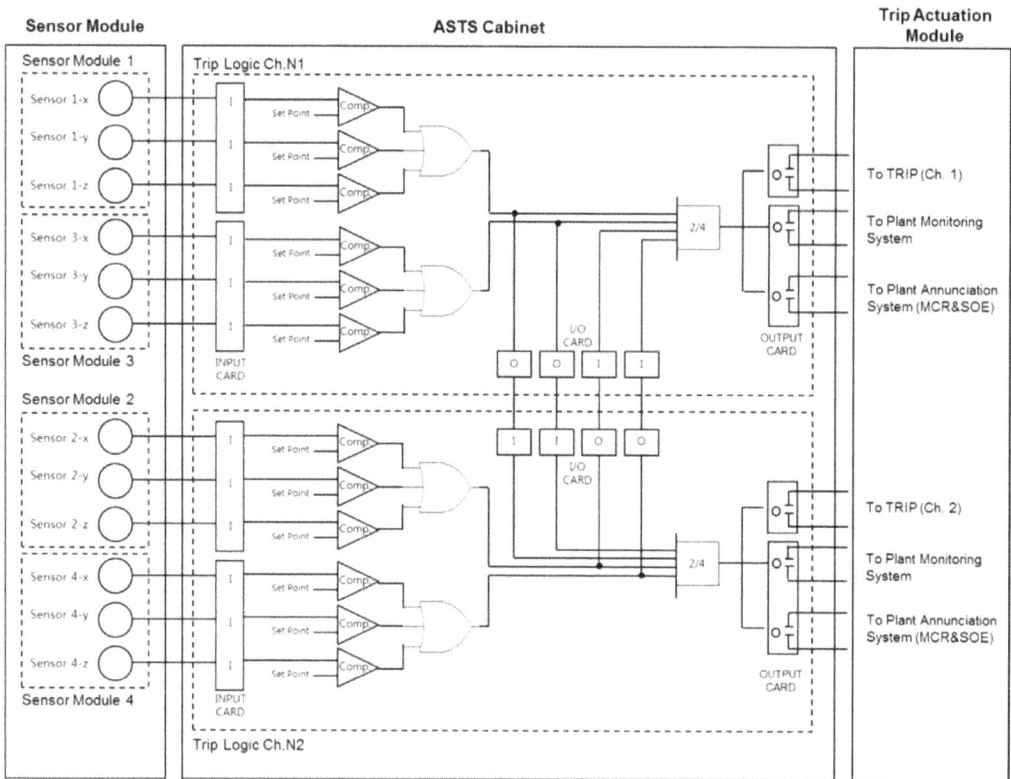

Figure 19. Overall scheme of digitalized ASTS [7].

The operational principles are as follows. When the measured signal exceeds the set-point, the edge triggering happens and the status is set to '0' by comparator actuation. In this case, the latch is engaged due to the edge triggering, and keep the status for 10 s. The purpose of 10-s latch is to synchronize the individual channel for the coincidence logic. If the signals are not properly synchronized, the trip initiation signal may not be actuated during a strong earthquake. ASTS consists of four diverse channel applications. The bi-stable signal from each channel feeds into two-out of-four coincidence logic for generating trip initiation signal.

5. Conclusions

In this work, two measurement principles are analysed to obtain the transfer function plotted by the frequency vs. relative response to the acceleration as well as the step responses. In order to define the sensitivity of open-loop type or closed-loop type with a feedback loop, two sensor types have been used; the force-balanced acceleration and the servo acceleration. The open-loop and closed-loop response when the step input describes the transfer function by plotting it with amplitude and time. The open-loop geophone shows exponentially decreasing function while the force balanced type illustrates overshoot before it is stabilized.

Two application examples for the NPP are discussed with their designs in this work. They are SMS and ASTS. When the Gyeongju earthquake occurred, the main concern in the NPP, whether the peak ground acceleration exceeded in OBE or SSE, was highlighted. The second issue was the frequency range determining the manual or automatic reactor trip. ASTS eliminates frequency over 10 Hz for generating a trip signal. Through Gyeongju earthquake the adequacy of this theory has been proved. The third issue that arose in ASTS design was a latching time of 10 s to synchronize the individual channel for the coincidence logic. This issue needs to be resolved by analysing the seismic waves in Gyeongju.

Author details

Jae Cheon Jung

Address all correspondence to: jcjung@kings.ac.kr

KEPCO International Nuclear Graduate School, Seosang-myeon, Ulju-gun, Ulsan, Republic of Korea

References

[1] "Strongest-ever earthquake hits Korea, tremors felt nationwide". The Korea Times. September 12, 2016. [Accessed 2016. 10].

[2] Katsuhiko Ogata, "Modern Control Engineering", 3rd ed. Chapter 3, New Jersey: Prentice-Hall; 1997. pp. 83–84

[3] Dept. of earth and environmental, Ludwig Maximilians Universiat https://www.geophysik.uni-muenchen.de/.../2_ms_seismic_instruments.pdf [Accessed 2016. 08]

[4] Kinemetrics FBA-23 User Manual, https://nees.org/data/get/facility/sensorModels/180/Documentation/FBA23.pdf [Accessed 2016. 08]

[5] Input/Output, SM-6 Geophone, http://www.iongeo.com/content/documents/.../DS_SEN_121026SM6.pdf [Accessed 2016. 08]

[6] GeoSig, AC-23 User Manual, http://www.geosig.com/files/GS_AC-23_UserManual_V12.pdf [Accessed 2016. 08]

[7] JC. Jung, Design of the Digitalized Automatic Seismic Trip System for Nuclear Power Plants, Nuclear Engineering and Technology, 46(2), 2014, pp. 235–246, http://dx.doi.org/10.5516/NET.04.2013.041;

7

Three-Dimensional Nepal Earthquake Displacement Using Hybrid Genetic Algorithm Phase Unwrapping from Sentinel-1A Satellite

Maged Marghany and Shattri Mansor

Additional information is available at the end of the chapter

Abstract

Introduction: Geophysicists had forewarned for decades that Nepal was exposed to a deadly earthquake, exceptionally despite its geology, urbanization and architecture. Gorkha earthquake is the most horrible natural disaster to crash into Nepal since the 1934 Nepal-Bihar earthquake. Gorkha earthquake occurred on April 25, 2015, at 11:56 NST and killed more than 10,000 people and injured more than 23,000 population. **Objective**: The main objective of this work is to utilize hybrid genetic algorithm for three-dimensional phase unwrapping of Nepal earthquake displacement using Sentinel-1A satellite. The three-dimensional best-path avoiding singularity loops (3DBPASL) algorithm was implemented to perform 3D Sentinel-1A satellite phase unwrapping. The hybrid genetic algorithm (HGA) was used to achieve 3DBPASL phase matching. Advancely, the errors in phase decorrelation were reduced by optimization of 3DBPASL using HGA. **Results**: The findings indicate a few cm of ground deformation and vertical northern of Kathmandu. Approximately, an area of 12,000 km^2 has been drifted also the northern of Kathmandu. Further, each fringe of colour represents about 2.5 cm of deformation. The large amount of fringes indicates a large deformation pattern with ground motions of 3 m. **Conclusion**: In conclusion, HGA can be used to produce accurate 3D quake deformation using Sentinel-1A satellite.

Keywords: Sentinel-1A satellite, Nepal earthquake, hybrid genetic algorithm, three-dimensional best-path avoiding singularity loops, decorrelation, interferogram

1. Introduction

Satellite-based interferometric synthetic aperture radar (InSAR) is a potential tool for precise measurements of ground shifts triggered off by earthquakes or landslides [1–4]. Whilst a satellite sweeps over a territory, it can pick up the amplitude and phase of radar pulses back-scatter off Earth's surface [5–7]. Subtracting the phases logged in different passes over the same region acquiesces an interference pattern which is sensitive to changes in topography along the satellite's line of sight within few millimetres [6, 8]. In other words, interferometric synthetic aperture radar (InSAR) techniques have excellent potentials to measure the millimetre scale of the Earth's surface deformation. Nevertheless, the foremost difficult raised up in InSAR techniques is phase unwrapping [8, 9].

The keystone questions are how the earthquake can generate based on plate tectonic theory and how InSAR can be used to detect earthquake deformation. The following two sections are devoted to understand the mechanisms of earthquake and InSAR technology. In this paper, we address the question of utilization of inversion genetic algorithm (GA) as an optimization methodused for three-diensional modelling of rate deformation due to Nepal earthquake displacement. In this context, Lee [10] used three-dimensional sorting reliabilities algorithm (3D-SRA) phase unwrapping for modelling volume rate changes of shoreline. However, 3D-SRA was not able to remove the artefacts in digital elevation model (DEM) due to radar shadow, layover, multipath effects and image misregistration and finally the signal-to-noise ratio (SNR) [11–16]. In fact, 3D-SRA does not identify singularity loops at all. It be determined by completely a quality measure to unwrap the phase volume. Ignoring singularity loops may cause the unwrapping path to penetrate these loops, and errors may propagate in the unwrapped phase map [17].

The two-dimensional unwrapping approaches, however, could introduce discontinuous regions when the noise is high. The resulting inconsistent baselines within a slice would produce an incorrectly unwrapped baseline. Then, the one-dimensional baseline unwrapping could give incorrect results. Many of the methods apply to quality map to guide the unwrapping procedures. The quality map was defined with the quality of the edges that connect two neighbouring voxels and unwrap the most reliable voxels first [18, 19]. Therefore, three-dimensional phase unwrapping approach considers the temporal domain and the spatial domain restrictions simultaneously [17, 18]. The main contribution of this study is to combine hybrid genetic algorithm (HGA) with three-dimensional phase unwrapping algorithm of three-dimensional best-path avoiding singularity loops (3DBPASL) algorithm with InSAR technique. Two hypotheses examined are (i) the HGA algorithm that can be used as filtering technique to reduce noise in the 3D phase unwrapping and (ii) 3D Nepal earthquake displacement that can be reconstructed using satisfactory phase unwrapping of 3DBPASL by involving HGA optimization algorithm.

2. Mechanism of earthquake based on plate tectonic theory

The main question that can arise is how do earthquakes generate? The answer of the above question required a comprehensive understanding of the theory of plate tectonic. Truthfully, plate tectonic theory is a starting point to capture the mechanisms of tsunami generation. Consequently, the plate tectonic theory let us have inclusive thought about generation of earthquakes and volcano which are the foremost causes for tsunami.

Consequently, there are three versions for the plate tectonic theory. The first version has discussed the lithosphere broken into strong, rigid moving plates that carry the continents on their backs ocean basins that come and go at mid-ocean ridges and subduction zones, respectively (**Figure 1**) [20].

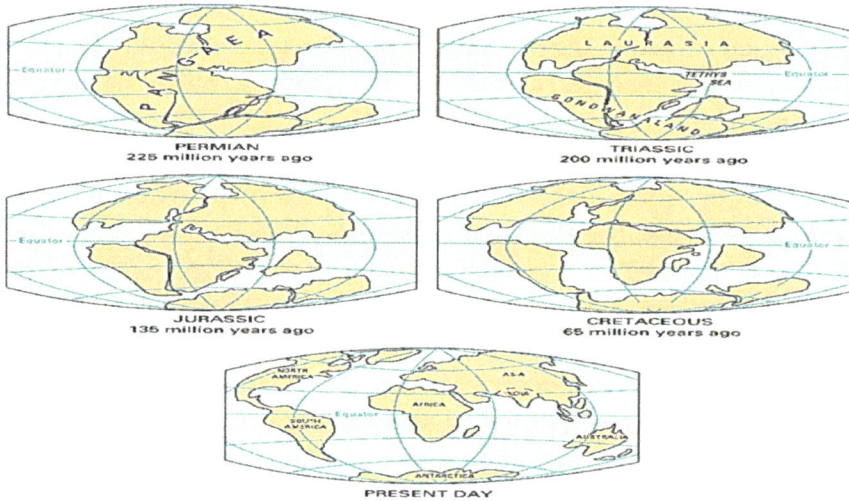

Figure 1. First version of plate tectonic theory.

Further, Marghany [21] has accepted that physical plate margins fluctuate significantly sideways belt (i.e. beginning location to location sideways their dimensions), combined plate margin kinds are usual. Therefore, plate margins are not completely unbending Successively and an entirely plate exchanges would correspondingly explicate, otherwise by slightest crack through, huge interior features similar mountain stripes and extensive faults intraplate distortions alike the rock-strewn mountains and washbasin. For instance, Hawaiian Islands represent the foremost mid-plate volcanic disruptions (**Figure 2**) [21]. This agrees with the study conducted by Marghany [20].

Marghany [22] stated that the major feature of plate tectonics is the lithosphere. With this regard, it is considered as unstable thermodynamic system which is staging to disintegrate the Earth's crust and mantle thermal. Further, Nitecki et al. [23] reported that plates are dim in expansion and penetrable to liquefy from the original, hardly dense asthenosphere. The absolute dynamic influences ahead of plate movements are the vertical cooling fluctuations and gravity force. The lower mantle is heating the upper mantle through conductivity. Further, 60% of seafloors and subduction cool the lower mantle. Consequently, subduction which is correlated to joint rollback and overruling plate leeway which have disappeared, for instance, the essential profiles of the Earth's surface. Extreme volcanism is considered as hot spots which concentrated by expansion of plate weakness. Besides, hot spots can be caused by limited and comparatively shallow thermal instabilities which are associated to thermal fluctuation of the upper mantle because of dynamic motions of adjacent plates [22].

Understanding the main features of plate tectonics can lead to extremely perceptive of earthquake occurrence mechanisms. According to Marghany [21], plate tectonics are associated by major features which involved:

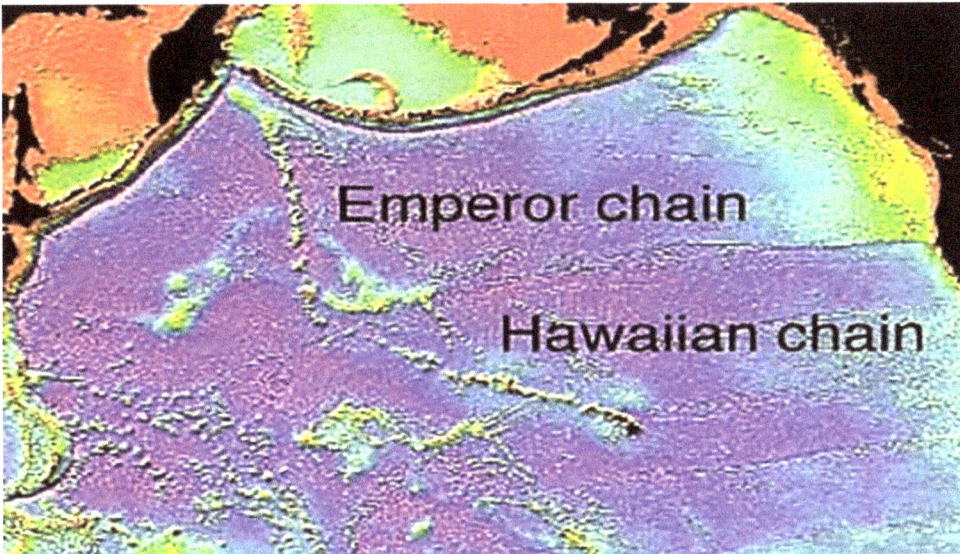

Figure 2. Northern Pacific floor with Hawaii-Emperor and many other volcanic chains.

- The surface of Earth is constructed of chains of huge plates.

- Per year, these plates are in steady movement drifting at a limited centimetres.

- The ocean bottoms are frequently shifting, scattering starting the midpoint and descending on the boundaries.

- The plates are moving in various paths due to the effects of convection current below the plates.

Consistent with above prospective, the radioactive decay is the main source of thermal fluctuations which is creating the convection current flows. This occurs deeply in the Earth which is deriving vertical density exchanges of fluid through the Earth. With this regard, less dense of hot rocks below the mantle rises above cooler rock which emerged down due to the impact of gravity on their high densities (**Figure 3**) [21].

The comprehensive understanding of tectonic plate interactions is keystone of earthquakes, volcano and tsunami generations. With this regard, the principles of tectonic plate interactions are (i) divergent boundaries, (ii) convergent boundaries and (iii) transform boundaries (**Figure 4**) [20].

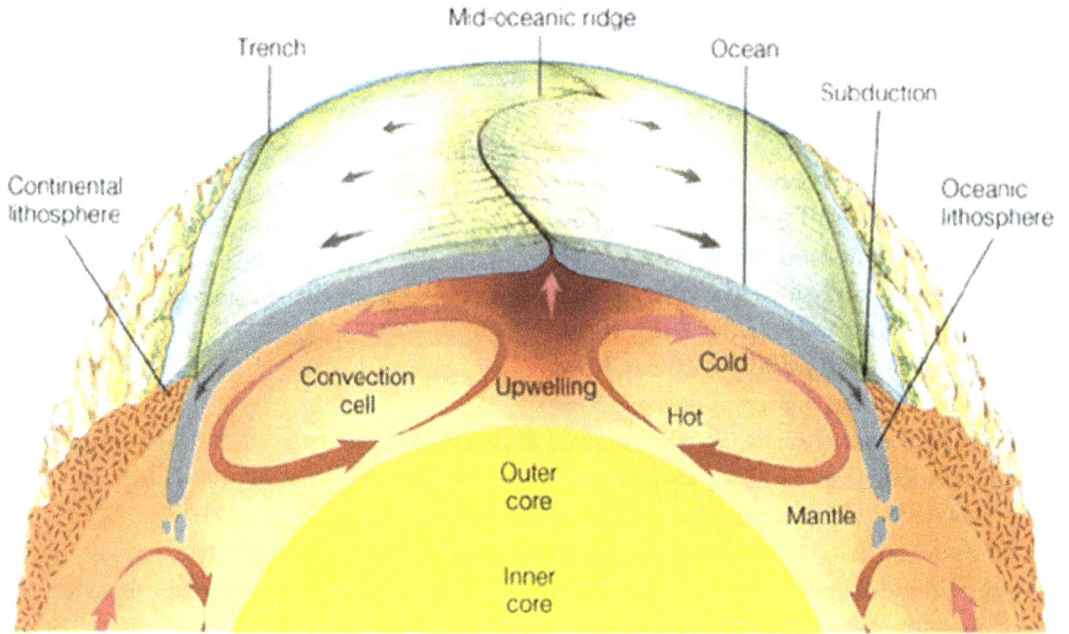

Figure 3. The convection currents of the Earth.

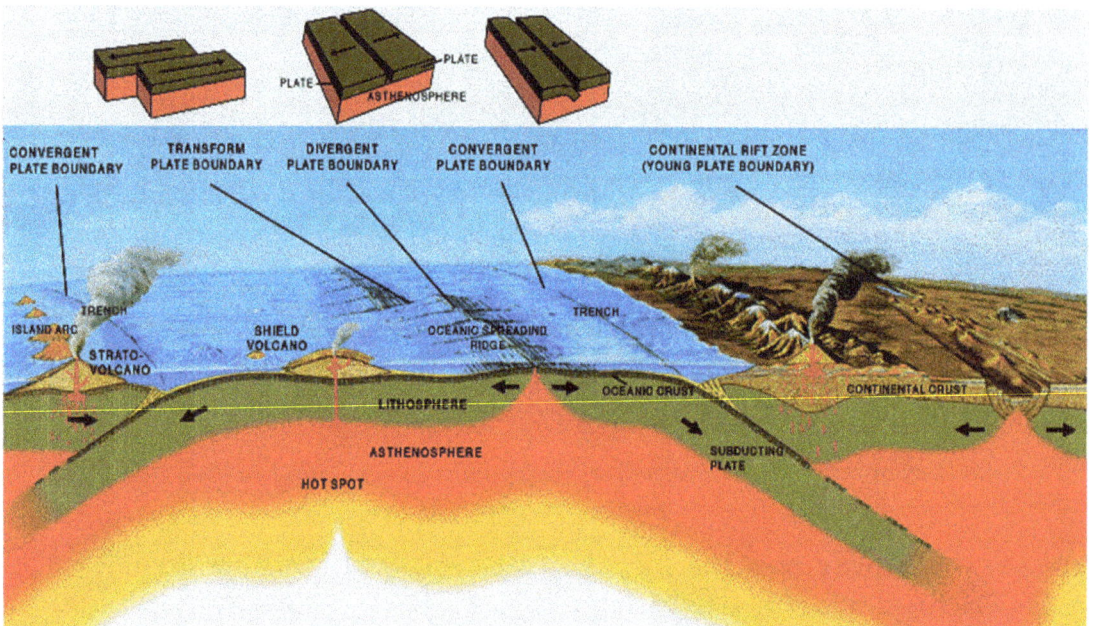

Figure 4 Three types of plate boundary.

Consequently, divergent boundaries represent zones where plates deviate from each other which are forming, for instance, (**Figure 5**) either mid-oceanic ridges or rift valleys. With these regards, divergent boundaries are also identified as constructive boundaries [20]. The Atlantic

Ocean was created by this process. The Mid-Atlantic Ridge is an area where a new seafloor is being created [24–26].

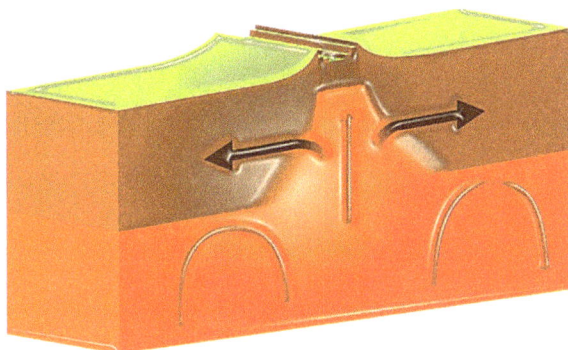

Figure 5. Divergent boundaries.

Convergent boundaries are well known as compressional or destructive margins. Along these boundaries, the plates approach each other and crash (**Figure 6**). These boundaries involved (i) subduction zones, (ii) obduction and (iii) orogenic belts. Subduction zones arise where an oceanic plate convenes a continental plate and is broke underneath it. Subduction zones are indicated by oceanic trenches (**Figure 4**). The descendent completion of the oceanic plate melts down and constructs compression in the mantle, triggering volcanoes to create.

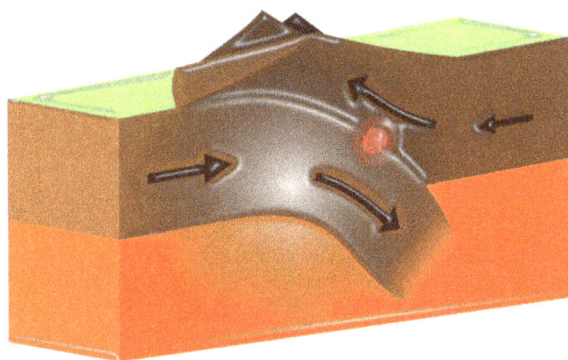

Figure 6. Convergent boundaries.

According to Marghany [24], the continental plate can make obduction arises up as soon as it is plugged underneath the oceanic crust. Nonetheless, this phenomena is considered unfamiliar, for instance, the plate tectonics have absolute densities which enable the oceanic crust's subduction. This heritage, the oceanic crust to fasten and frequently upshot in a different fangled mid-ocean ridge, for instance, is evolving and spinning the seizure hooked on subduction [25–28].

Consequently, the collision boundaries are known as orogenic belts [24, 28]. These collision boundaries are produced because of collision of two plates which thrust upwards to create bulky mountain ranges, for instance, the Himalayas, the highest mountain range on Earth (**Figure 7**).

Figure 7. Mountain ranges of the Himalayas.

Furthermore, Marghany [21, 24] and Ferraiuolo et al. [27] agreed that the most spectacular geological feature on the surface of the Earth is orogen restraint. This can be seen between crust boundaries of Indo-Australian crust and African crust and to the north of the Eurasian crust. This restraint runs from beginning of New Zealand, cutting edge of the East-South-East, from side to side of Indonesia, sideways of the Himalayas, from end to end of the Middle East and moves up to the Mediterranean in the West-Northwest. It is also termed the 'Tethyan' District, as it creates the zone lengthways which the earliest Tethys Ocean was malformed and vanished [25].

Figure 8 shows occurrence of transform boundaries when two plates suppress past each other with only partial convergent or divergent movement.

Figure 8. Transform boundaries.

The San Andreas Fault in California, for instance, one of the well-known transform boundaries, is an active transform boundary. The Pacific Plate (carrying the city of Los Angeles) is moving northwards with respect to the North American Plate [26].

The critical question is how to prove the occurrence of plate tectonics? The landmasses appear to be appropriate and organized, similar a massive picture puzzle. In this respect, when anyone goes into a map, Africa appears to nestle agreeably addicted to the Caribbean Sea and the east coast of South America. In line with Ferraiuolo et al., [27], in 1912, Alfred Wegener suggested that these two landmasses were as soon as combined with each other formerly someway moved away from each other. He identified this procedure as Pangea. He assumed that Pangea was together up to 200 million years ago. The map of plate tectonics has been established by NASA and is shown in **Figure 9** which has an excellent evidence of the existence of plate tectonics [27].

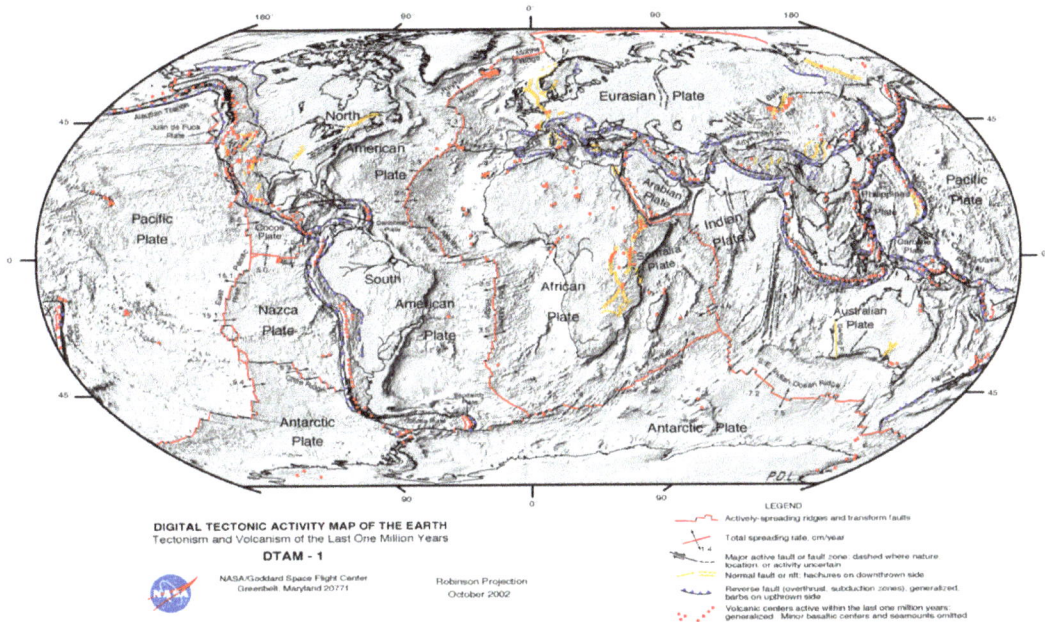

Figure 9. Current map of plate tectonics.

3. Continental drift

In 1915, the theory of continental drift was proposed by the German geologist and meteorologist Alfred Wegener, which declares that parts of the Earth's crust slowly drift atop a liquid core. The fossil record verifies and provides credibility to the theories of continental drift and plate tectonics. The theory of continental drift has been counted by the theory of plate tectonics, which clarifies how the continents shift (**Figure 10**). The impression that continents can drift about is termed, not surprisingly, continental drift. The old (and very strong!!) theory earlier was the 'contraction theory' which recommended that the Earth was once a molten ball and in the process of cooling the surface cracked and folded up on itself [28].

The critical issue with this clue was that all mountain ranges should be roughly of the same age, and this cannot be proper. It was suggested that as the continents relocated, the primary edge of the continent would confront resistance and thus constrict and bend upwards creating mountains close to the foremost edges of the drifting continents. Consistent with Ferretti et al.,

[28], there is contrariety that the key protest to continent drift was that there is no mechanism, and plate tectonics was recognized without a mechanism, to transfer the continents.

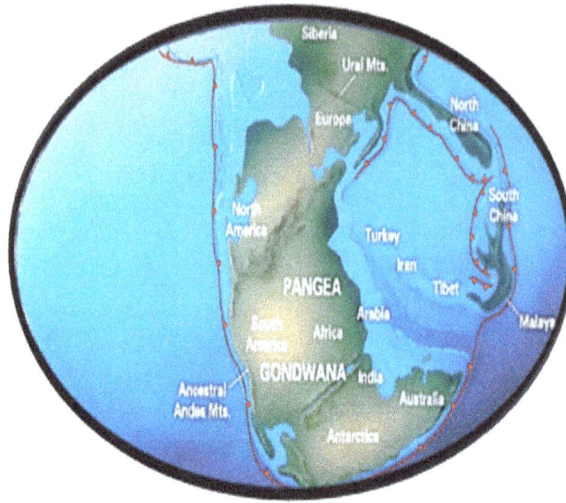

Figure 10. Continental drift.

However, plate tectonics are the extensively conventional theory that Earth's crust is broken into rigid, transferring plates. In the 1950s and 1960s, scientists revealed that the plate edges through magnetic surveys of the ocean floor and through the seismic heeding networks assembled to examine the nuclear trying. Discontinuous patterns of magnetic anomalies on the ocean floor specified seafloor spreading. With this regard, a new plate material is born. Magnetic minerals affiliated in ancient rocks on continents correspondingly exposed that the continents have lifted completely to one another [25–28]. For instance, India drifted northwards into the Asia forming the Himalayas and of course Mount Everest (**Figure 11**).

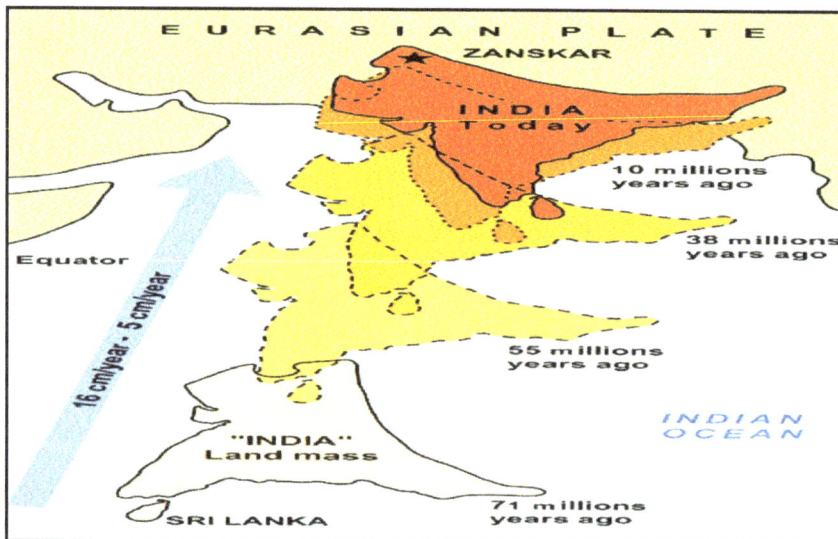

Figure 11. Himalayas theory of drifting tectonic.

The East Asian Sea Plate was an unknown tectonic plate which has been swallowed up by the Earth. It can be found in the Philippine Sea. In fact, the Philippine Sea locates at the juncture of numerous foremost plate tectonics: (i) the Pacific, (ii) Indo-Australian and (iii) Eurasian plates. These set up numerous minor plates, containing the Philippine Sea Plate. Since approximately 55 million years ago, the Philippine Sea Plate has been drifting northwest [27].

4. Nepal earthquake

Gorkha earthquake is the most horrible natural disaster to crash into Nepal since the 1934 Nepal-Bihar earthquake. Gorkha earthquake occurred on April 25, 2015, at 11:56 NST, killed more than 10,000 people and injured more than 23,000 population. Its epicentre was the east of the District of Lamjung, and its hypocentre was with an approximately depth of 15 km with maximum magnitude of 8.1 M_w. Consequently, within 15–20 min, aftershock was struck across Nepal with a magnitude of 6.7 on the 26th of April at 12:54:08 NST. Thus, the epicentre of a foremost aftershock was close to the Chinese border between the capital of Kathmandu and Mount Everest (**Figure 12**) with a moment magnitude of 7.3 M_w [29].

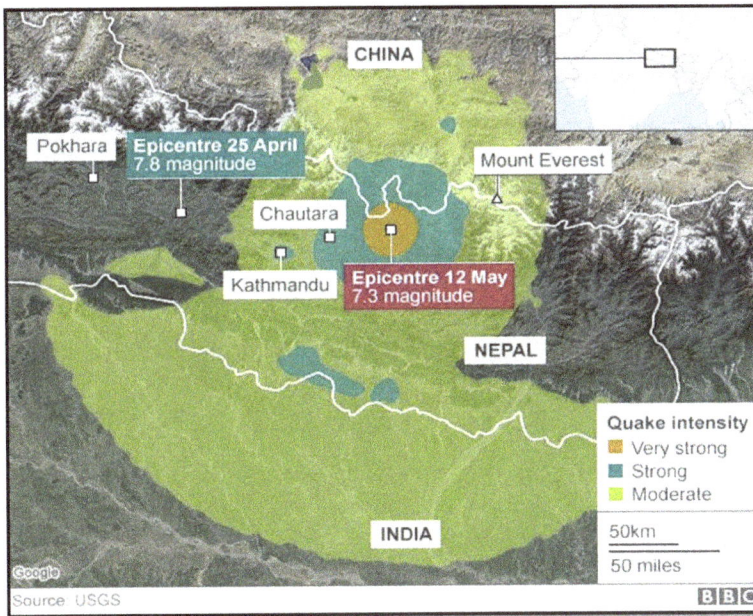

Figure 12. Location of Nepal's earthquake.

Consistent with the USGS [30], the temblor was triggered through an abrupt push. Further, the accumulative energy underneath the surface of the Earth produced huge stress on the main fault which is associated with the Indian crust. This was contributed to drift the Indian crust further down the Eurasian crust. With this regard, Kathmandu sited on a slab of layer almost 120 km widespread and 60 km elongated. Within 30 s, the crust was drifted 300 cm southward.

Counter to Hussein et al. [31], Nepal situated southwards which is closed to the slab crust somewhere the Indian crust moved down of the Eurasian crust. This was the plate dominating the essential locality (**Figure 13**) within 800 km of the Himalayan arch. Organically, the Nepal Himalayas are segregated into five tectonic district morpho-geotectonic zones: (1) Terai valley, (2) Sub-Himalaya, (3) Lesser Himalaya (Mahabharat Range and mid-valleys), (4) Higher Himalaya and (5) Inner Himalaya (Tibetan-Tethys) [32, 59, 60, 61]. In central Nepal, the convergence rate between the plates is about 45 mm/year. According to USGS [30], the location, magnitude and focal mechanism of the earthquake indicate that it was triggered off by a slip along the Main Frontal Thrust (**Figure 14**).

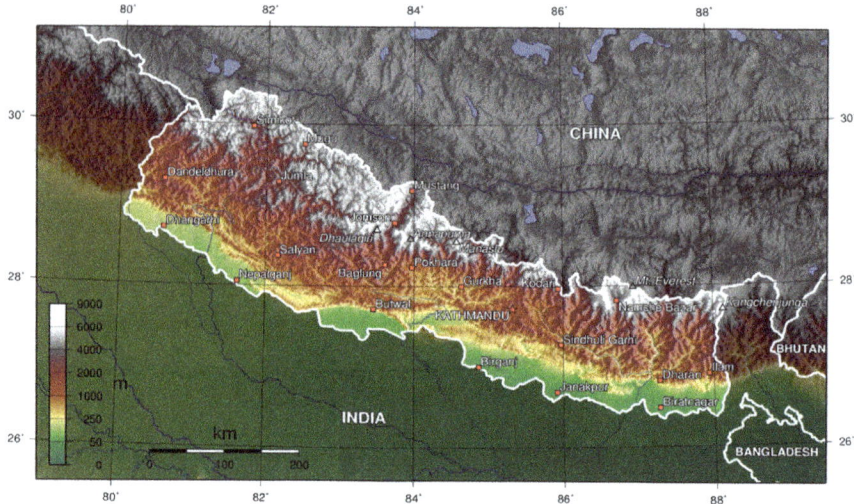

Figure 13. Topography map of Nepal.

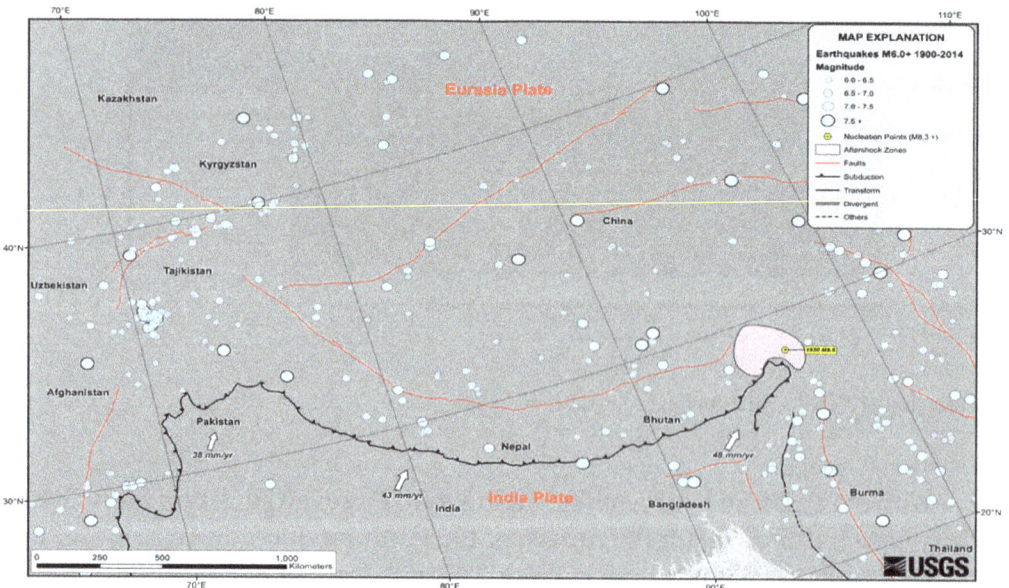

Figure 14. Nepal Himalayas are partitioned into five tectonic zones.

The earthquake's effects were amplified in Kathmandu as it sits on the Kathmandu Basin, which contains up to 600 m of sedimentary rocks, representing the infilling of a lake. Wei and Cumming [33] stated that the latest quake follows the same pattern as a duo of big tremors that occurred over 700 years ago and results from a domino effect of strain transferring along the fault. The last time the fault ruptured at this location was back in 1344. It was preceded in 1255 by a big event to the east of Kathmandu. The last rupture there was in 1934, hinting strain might accumulate westwards. This means that 2015's quake follows the pattern with a gap between events of 80 years or so. In other words, the Main Frontal Thrust, on average of a great earthquake, occurs every 750 ± 140 and 870 ± 350 years in the east Nepal region. Perhaps, the 700-years delay between earthquakes in the region. Finally, it can be suggested that the 2015 quake was due to the tectonic stress buildup from the 1934 quake [32].

5. Interferometric synthetic aperture radar for earthquake monitoring

The SAR interferometry is established on the SAR machinery. The abbreviation SAR tolerates for synthetic aperture radar. SAR mechanism is a function of signal transmission-reception of a percentage of energy that interrelates with the surface, which is denoted as backscattered, existing a determining strength and time delay of the received signals [34].

Consistent with Massonnet et al. [1], InSAR is a skeleton covering various approaches or procedures that expend phase knowledge obtained by monitoring phase variance or status of displacement of the microwave at the instantaneous which is obtained by the radar antennas between two SAR imageries. With this regard, it is recognized as master and slave images. Therefore, the interferometric phase is assimilated from diverse sensor locations [1, 3, 5, 35, 36]. In this concern, map variations of digital elevation model (DEM) can be created by employing the two single-look complex (SLC) synthetic aperture radar (SAR) data which are obtained by two or extra disconnect antennas. As stated by Qifeng et al. [37], the output phase scene's result is generated by multiplying the coregistered complex conjugate pixels and complex radar data. By the way, the difference of the two SAR phase data is handled to obtain DEM and/or the earth's surface distortion. Consistent with Qifeng et al. [37] and Nan et al. [36], the absolute value of the real phase difference between any two neighbouring pixels is less than π, and then the real phase can be acquired by integrating the wrapped values of the wrapped phase differences. Nevertheless, wrapped phase image is extremely challenged by several factors. Particularly, these critical factors are involved multiplicative speckle noises, shadow, foreshortening, layover, temporal, geometric and atmospheric decorrelations [1, 3, 37–39] which are negatively produced in the area of nonstandard phase, i.e. low-quality area [36, 40, 41]. In this respect, low-quality area can contribute to critical decorrelation issues in the phase unwrapping procedures.

Several phase unwrapping algorithms have been introduced to solve the critical issue of low-quality area and the decorrelation. These algorithms are categorized into (i) path-following algorithms and (ii) minimum-norm algorithms [42]. Subsequently, minimum-norm algorithms express the unwrapping issue in terms of minimization of the global function as compared to path-following algorithms. Conversely, the constraint of minimum-norm algorithms cannot be involved every individual pixel in the phase unwrapping procedures [63]. In contrast, path-following algorithms are extremely advanced compared to minimum-norm algorithms. The advances of path-following algorithms are (i) to identify the residues and use the quality map to guide the generation of branch cuts by implementing branch-cut [41] and mask-cut algorithm [43], (ii) to guide the path of integration which is function of quality-guided algorithm and (iii) to dismiss the total discontinuities of the unwrapped result by using minimum discontinuity algorithm [44, 45]. Besides, there are other phase unwrapping algorithms. Therefore, the conventional phases unwrapping region-growing and the least-square algorithms require accurate image coregistration, which should be about 1/10 to 1100 resolution cell size (i.e. SAR pixel) [46]. In this context, Hai and Renbiao [5] stated that the interferometric phase unwrapping method which is based on subspace projection can provide an accurate estimation of the terrain interferometric phase (interferogram) even if the coregistration error reaches one pixel. This can be achieved by phase unwrapping algorithms such as the branch-cut method [11, 46].

Commonly, the accuracy of the image coregistration is a serious issue for accurate interferometric phase unwrapping. It is well known that the performance of interferometric phase unwrapping suffers seriously from poor image coregistration. Therefore, TanDEM-X and TerraSAR-X have been implemented to maintain the issue of image coregistration. Further, ERS-1 and ERS-2 and TerraSAR-X in tandem mode are the excellent examples of short temporal resolution. In wide range of contexts, TanDEM-X and TerraSAR-X are imaging the terrain below them simultaneously, from different angles. These images are processed into precise elevation maps with a 12 m resolution and any vertical accuracy better than 2 m [1, 3, 15, 39, 46, 47].

Further, Gens [14] described that the period change of two SAR acquirements is an additional kind of decorrelation. Certainly, the period modifications despite the fact associate with SAR data groups through a comparable dimension of baseline which attained one and 35 days which causes temporal component decorrelation. Therefore, Dickinson [48] stated that the loss of coherence in the same repeat cycle in data acquisition is most likely because of baseline decorrelation and dense vegetation covers in such a tropical as Malaysia. According to Nizalapur et al. [7], uncertainties could arise in DEM [14] because of the limitation of the InSAR repeat passes [3, 9, 15, 23, 36, 49, 56, 58, 62]. A persistent phase alteration throughout the two SAR data which is produced by the straight regular atmosphere was above the dimension measure of an interferogram and vertically above the landscape. The atmosphere, nevertheless, is alongside dissimilar on dimension measures equally greater and lesser than usual distortion signals [8, 23].

Recently, Zhong [44] stated that Differential Interferometric Synthetic Aperture Radar (DInSAR) has recently been applied with success to investigate the temporal evolution of the detected deformation phenomena through the generation of displacement time series. With this regard, two foremost kinds of progressive DInSAR practices of distortion period-chain origination have been recommended in several scientific works, which are frequently mentioned as persistent scatterer (PS) and small baseline (SB) procedures, correspondingly. The PS algorithms choose completely the interferometric SAR data pairs through indication to a single SAR master scene, deprived of slightly restraint on the sequential and three-dimensional parting (baseline) amongst the orbits.

6. Phase unwrapping

Phase unwrapping is a key step in modelling surface deformation from interferometric synthetic aperture radar (InSAR) data. Thus, the measured phases are wrapped in the interval $(2\pi, \pi]$, and phase unwrapping is a necessary step to recover the original full phase value [4]. In particular, phase unwrapping algorithms can be classified into two categories: (i) the path-following method and the (ii) minimum-norm approach. Goldstein's method is a classic path-following phase unwrapping method which explores for excesses cutting edge of the phase scene and employs method of branch cuts amongst closed interferogram errors. In this manner, it eliminates the mixing route dependency once the phase scene is perfectly unwrapped. The phase alteration mixing departs from the beginning of the upper leftwards corner of the scene. It followed by unwrapping the phases through the hypothesis of that contiguous pixels which have phase variances contained by the array of $[-\pi, \pi]$. Similarly, the unwrapping route requests to assimilate branch cuts with the intention of eliminating the route dependence. Nevertheless, after the branch cuts are positioned in the incorrect site, the wrong unwrapping is produced [4].

The minimum-norm approaches usually require extra calculation period than the route-succeeding mixing procedures. In Flynn's algorithms, for instance, Gao and Yin [46] attempt to diminish the cutoff in phase results. From the algorithm it was discovered that the pathway can use the loop of incoherence and can eliminate the gap by toting or eliminating 2π on the phase standards inside the ring section. The procedure is achieved by pending to no extra incoherence loops which are created. In this respect, the incoherence placements can be directed by using a quality map [40].

In proportion to ENVISAT and Karout [19, 40], the keystone of the quality map concept is firmly based on the approach which unwraps laterally a route initial through the utmost excellence pixel. With this regard, slight unwrapping is compulsory to the furthermost vague spaces everywhere the phases of contiguous recovered superiority pixels are implemented to achieve precisely unwrapping levels. The quality map can be a correlation map, phase derivative variance map, etc. When a correlation map is not available, the phase derivative variance map is generally used. In the word of Mitri [50], the phase estimation is a major challenge to determine more accurate DEM. This is because of the measured phase differences which are given as a wrapped phase field of the principal values of a range $-\pi$ to π; thus, the existence is unspecified within multiples of 2π [7]. This procedure produces phase jumps between neighbouring pixels. Smooth function is used to resolve phase jump by adding or subtracting multiples of 2π. Consequently, multichannel MAP height estimator based on a Gaussian Markov random field (GMRF) has developed by Wikipedia [51] and Frankel [52] to solve the uncertainties of DEM reconstruction from InSAR technique. They found that the multichannel MAP height estimator has managed the phase discontinuities and improved the DEM profile [53]. Recently, Bollinger et al. [54] validated GMRF technique with a lower range of error (0.01 ± 0.11 m) with 90% confidence intervals for DEM reconstruction using RADARSAT-1 SAR F1 mode data.

However, the two-dimensional unwrapping methods could introduce discontinuous regions when the noise is high. The resulting inconsistent baselines within a slice would produce an incorrectly unwrapped baseline. Then, the one-dimensional baseline unwrapping could give incorrect results. Many of the methods are applied to quality map to guide the unwrapping procedures. The quality map was defined with the quality of the edges that connect two neighbouring voxels and unwrap the most reliable voxels first [31, 32]. Therefore, three-dimensional phase unwrapping approach considers the temporal domain and the spatial domain restrictions simultaneously [10].

7. Deformation reconstruction using three-dimensional best-path avoiding singularity loops (3DBPASL) algorithm

Three methods are involved to perform InSAR from Sentinel-1A satellite data: (i) conventional InSAR procedures, (ii) three-dimensional phase unwrapping algorithm, i.e. three-dimensional best-path avoiding singularity loops (3DBPASL) algorithm [19, 40, 55], and [iii] hybrid genetic algorithm. Consistent with Zebker et al. [4], SAR interferometry (equivalently, the InSAR technique) is a technique to extract surface's physical properties by using the complex correlation coefficient of two SAR signals. The complex correlation coefficient, γ, of the two SAR observations, s_1 and s_2, is defined as

$$\gamma = \frac{E\left[s_1 s_2^*\right]}{\sqrt{E\left[s_1 s_1^*\right] E\left[s_2 s_2^*\right]}} \tag{1}$$

where $E[]$ is the mathematical expectation (ensemble averaging) and * represents the complex conjugate [2]. Further, Hanssen [3] stated that the interferometric phase is defined as the phase of the complex correlation coefficient as

$$\varphi = \arg\{\gamma\} = \arg\left\{E\left[s_1 s_2^*\right]\right\}, \tag{2}$$

and its two-dimensional map is called the interferogram. Moreover, Hanssen [3] and Zebker et al. [4] defined the coherence as the amplitude of the complex correlation coefficient which expressed as

$$\rho = |\gamma|, \tag{3}$$

and its two-dimensional map is presented in the coherence image [1, 3, 5]. According to Qifeng et al. [37], an interferogram contains the interferometric phase fringes from SAR geometry, together with those from topography and displacement of the surface. The level of coherence can give a measure of the quality of the interferogram. Initially, the InSAR techniques were mainly dedicated to topographic information retrieval from interferograms. Further development resulted in techniques to extract interferometric phase fringes from coherent block displacement of the surface [26, 51–53]. In Lee [10], the surface displacement can estimate using the acquisition times of two SAR data s_1 and s_2. The component of surface displacement, thus, in the radar-look direction, contributes to the further interferometric phase (ϕ) as [8, 12]

$$\phi = \frac{4\pi(\Delta R)}{\lambda} = \frac{4\pi(B_h \sin\theta - B_v \cos\theta)}{\lambda} \tag{4}$$

where ΔR is the slant range difference from satellite to target, respectively, at different times, θ is the look angle (20–46°) [19, 55], λ is Sentinel-1A satellite wavelength single-look complex (SLC) of C-band. Therefore, B_h and B_v are horizontal and vertical baseline components [10]. Then, the phase difference $\Delta\phi$ between the two Sentinel-1A data positions and the pixel of target of terrain point is given by

$$\Delta\zeta = \frac{\lambda R \sin\theta}{4\pi B} \Delta\phi \tag{5}$$

Eq. (5) is a role of standard baseline B and the range R. Furthermore, Eq. (5) can deliver evidence approximately of DEM and phase variance approximations. In truth, the predictable DEM of SAR images is a significant mission to construct the 3D distortion. However, this equation required short baseline and accurate image coregistration to acquire accurate quality of interferogram phase map. Phase unwrapping in Eq. (5) can be extended to three-dimensional to

$$\sum_{i,j,k} w^x_{i,j,k} \left| \Delta\phi^x_{i,j,k} - \Delta\psi^x_{i,j,k} \right|^P + \sum_{i,j,k} w^y_{i,j,k} \left| \Delta\phi^y_{i,j,k} - \Delta\psi^y_{i,j,k} \right|^P + \sum_{i,j,k} w^z_{i,j,k} \left| \Delta\phi^z_{i,j,k} - \Delta\psi^z_{i,j,k} \right|^P \tag{6}$$

where $\Delta\phi$ and $\Delta\psi$ are the unwrapped and wrapped phase differences in x, y and z, respectively, and w represents user-defined weights. The summations are carried out in between x, y and z directions over all i, j and k, respectively. A more advanced method developed by Haupt and Haupt [42] is L^P-norm which uses similar methods like the two previous least squares methods to solve the phase unwrapping problem. However, this method does not compute the minimum L^2-norm but the general minimum L^P-norm. In essence, by computing the minimum L^P-norm where $p \neq 2$, this method can generate data-dependent weight unlike the weighted least squares method. The data-dependent weights can eliminate iteratively the presence of the residues in the unwrapped solution. This method is more robust than the previous mentioned least squares method, and it is more computationally intensive [45]. Then the phase unwrapping based on the quality map can be calculated using the following equation [10, 40]

$$Q_{m,n,l} = \frac{1}{m \times n \times l} * \left(\left(\sum (\Delta\phi^x_{i,j,k} - \overline{\Delta\phi^x_{i,j,k}}) \right)^2 \right)^{0.5}$$
$$+ \left(\left(\sum (\Delta\phi^y_{i,j,k} - \overline{\Delta\phi^y_{i,j,k}}) \right)^2 \right)^{0.5} + \left(\left(\sum (\Delta\phi^z_{i,j,k} - \overline{\Delta\phi^z_{i,j,k}}) \right)^2 \right)^{0.5}, \tag{7}$$

where $\Delta\phi^x$, $\Delta\phi^y$, and $\Delta\phi^z$ are the unwrapped phase gradients in the x, y and z directions, respectively. $\overline{\Delta\phi}^x$, $\overline{\Delta\phi}^y$, and $\overline{\Delta\phi}^z$ are the mean of unwrapped phase gradient in $m \times n \times l$ cube in $\Delta\phi^x$, $\Delta\phi^y$, and $\Delta\phi^z$, respectively. i, j and k are neighbours' indices of the voxel v_in m, n and l cube. Following ENVISAT [19], the maximum gradient of the voxel v_m, n, l_ can be obtained by estimating the maximum unwrapped phase gradient of in the x, y or z directions, as described in Eq. (8):

$$X_{m,n,l} = \max\left\{ \max\left\{ |\Delta\,\phi^x_{i,j,k}| \right\}, \max\left\{ |\Delta\,\phi^y_{i,j,k}| \right\}, \max\left\{ |\Delta\,\phi^z_{i,j,k}| \right\} \right\} \qquad (8)$$

According to ENVISAT [19] the maximum gradient method indicates the badness rather than the goodness of the unwrapped phase data, so the quality is calculated using the reciprocal of the unwrapped phase gradient of Eq. (8). Further, ENVISAT and Karout [19, 40] and Lee [10] agreed that quality-guided phase unwrapping algorithms are the function of the quality of the voxels themselves to conduct the unwrapping path and to minimize error propagation during the unwrapping procedure. In this respect, the unwrapping path algorithm is the function of the quality of the edges as an intermediate stage, rather the quality of the voxels [19, 40]. Following Lee [10] and Karout [40], the quality map of voxels can be given by

$$Q_{m,n,l} = \sqrt{O_x^{\,2}(m,n,l) + E_y^{\,2}(m.n,l) + L^2(m.n,l)} \qquad (9)$$

where O_x, E_y and L are the horizontal, vertical and normal second differences, respectively [10], where

$$
\begin{aligned}
O_x(m,n,l) &= \Delta\varphi^x_{m,n,l}[\varphi^x_{m+1,n,l} - \varphi^x_{m,n,l}]^{-1}[\partial\varphi^x_{m-1,n,l} - \partial\varphi^x_{m,n,l}] \\
&\quad - d\varphi^x_{m,n,l}[\varphi^x_{m+1,n,l} - \varphi^x_{m,n,l}]^{-1}[\partial\varphi^x_{m,n,l} - \partial\varphi^x_{m+1,n,l}],
\end{aligned}
\qquad (10)
$$

$$
\begin{aligned}
E_y(m,n,l) &= \Delta\varphi^y_{m,n,l}[\varphi^y_{m,n+1,l} - \varphi^y_{m+1,n,l}]^{-1}[\partial\varphi^y_{m,n-1,l} - \partial\varphi^y_{m,n,l}] \\
&\quad - d\varphi^y_{m,n,l}[\varphi^y_{m,n+1,l} - \varphi^y_{m,n,l}]^{-1}[\partial\varphi^y_{m,n,l} - \partial\varphi^y_{m,n+1,l}],
\end{aligned}
\qquad (11)
$$

$$
\begin{aligned}
L(m,n,l) &= \Delta\varphi^z_{m,n,l}[\varphi^z_{m,n,l+1} - \varphi^z_{m,n,l}]^{-1}[\partial\varphi^z_{m,n,l-1} - \partial\varphi^z_{m,n,l}] \\
&\quad - d\varphi^z_{m,n,l}[\varphi^z_{m,n,l+1} - \varphi^z_{m,n,l}]^{-1}[\partial\varphi^z_{m,n,l} - \partial\varphi^z_{m,n,l+1}],
\end{aligned}
\qquad (12)
$$

where m, n and l are the neighbours' indices of the voxel in $3 \times 3 \times 3$ cube and $d\varphi^z_{m,n,l}[\varphi^z_{m,n,l+1} - \varphi^z_{m,n,l}]^{-1}$ defines an unwrapping operator that unwraps all values of its argument in the range $[-\pi, \pi]$. This can be done by adding or subtracting an integer number of 2π rad to its argument [10, 19, 51–53]. Eqs. (10)–(12) are signified 3D displays of the gradient in unwrapped phase variations of $d\varphi^x$, $d\varphi^y$, and $d\varphi^z$. Furthermore, the determined phase incline estimates the scale of the biggest phase slope which is fractional derived or draped of the phase alteration in V^i, V^j, and V^k dimensions [40].

In general, the quality of separately border of phase unwrapping is a function of the connection of two voxels in 3D Cartesian axis, e.g. x, y and z. Starting to optimize the unwrapping path from high-quality voxels to bad quality voxels [19] by using hybrid genetic algorithm (HGA). Indeed, the 3D best path which is avoiding singularity loops (3DBPASL) algorithm may not be the shortest when the residues are dense as the isolated districts enclosed by the branch cuts can easily appear. Searching for the shortest path in the path-following algorithm is in fact an optimization problem. Several optimization approaches, such as genetic algorithms, have been applied to 2D phase unwrapping. Therefore, hybrid genetic algorithm (HGA) has numerous advantages, for instance, global searching, robustness, parallelism. Further, it can be implemented with other algorithms.

8. Hybrid genetic algorithm

Following Ghiglia and Pritt [45], the HGA algorithm relies on estimating the parameters of an nth order of polynomial to approximate the unwrapped surface solution from the wrapped phase data. The coefficients of the polynomial that best unwrap the wrapped phase map are obtained by initial solution of GA algorithm to avoid a long time to converge to the global optimum solution. In this context, GA minimizes minimum 3DBPASL and $Q_{i,j,k}$ errors between the gradient of the polynomial unwrapped surface solution and the gradient of the original wrapped phase map. In other words, more precision and lower minimum 3DBPASL and $Q_{i,j,k}$ errors are achieved by increasing the order of the polynomial. This proposed algorithm is mainly applicable to adjoining phase distributions (albeit with gaps). Any optimization problem using a GA requires the problem to be coded into GA syntax form, which is the chromosome form. In this problem, the chromosome consists of a number of genes where every gene corresponds to a coefficient in the nth order of surface-fitting polynomial as described into Eq. (13):

$$f := n \ \rightarrow \ \sum_{k=0}^{n} \sum_{j=0}^{n} \sum_{i=0}^{n} a_{i,j,k} \Delta \phi_{i,j,k}^{x} \Delta \phi_{i,j,k}^{y} \Delta \phi_{i,j,k}^{z} \tag{13}$$

where $a[0.... n]$ is the parameter coefficients which are retrieved by the genetic algorithm to approximate the unwrapped phase that can achieve the minimum 3DBPASL and $Q_{i,j,k}$ errors. Further, i, j and k are indices of the pixel location in the unwrapped phase, respectively, and n is the number of coefficients.

The initial population is generated by creating an initial solution using one of the quality-guided phase unwrapping algorithms (3DBPASL algorithm) [40]. Following Ghiglia and Pritt [45], the initial solution is approximated using a 'polynomial surface-fitting-weighted least squares multiple regression' method. The initial population is then generated based on the initial solution. In doing so, in every a_g in each chromosome in the population, a small number relying on the accuracy of the gene is added or subtracted to the value of the gene as given by [44]

$$a_g = a_g + (\pm 1)\{10^{[\log(a_s) + \Re]}\} \tag{14}$$

where a_g is the coefficient parameter stored in gene g and R is a random number generated between the values.

Calculate the objective values of chromosomes in the population, and record the Pareto-optimal solutions.

Let Q_0, Q_1, Q_2 $\in$ P and P is an achievable locality. And, Q_0 is called the Pareto-optimal finding of Q_0 in the minimization problematic decision which is taken under the following satisfied circumstances.

1. If $f(Q_1)$ is supposed to be moderately larger than $f(Q_2)$, i.e. $f_m(Q_1) \geq f_m(Q_2)$, $\forall$ $i = 1, 2, \ldots, n$ and

$$f_m(Q_1) > f_i(Q_2), \exists\ i = 1, 2, \ldots, n, \tag{15}$$

then Q_1 is assumed to be subjugated by Q_2.

2. Suppose there is no $Q \in P$ St. Q dominates Q_0, then Q_0 is the optimal solution of Pareto.

In this step, the Q quality solution is evaluated at every generation to determine the global optimum solution to the parameter estimation-phase unwrapping problem. Therefore, the genes of a chosen chromosome are substituted as coefficients in Eq. (15) to evaluate the approximated phase value at coordinate (i, j, k). Then, the obtained phase is subtracted from the contiguous pixel approximated phase value to retrieve the approximated unwrapped phase solution gradient. It is then subtracted from the gradient of the wrapped phase in the i, j and k directions [39,44].

Following Hai and Renbiao [56], the two-point greedy continuous crossover is implemented in crossover operator. Therefore, crossover is less problem than the mutation operator. Thus, mutation operator concerns deliberate changes to a gene at random, to keep variation in genes and to increase the probability of not falling into a local minimum solution [41–44]. It involves exploring the search space for new better solutions. This proposed operator uses a greedy technique which ensures that only the best-fit chromosome is allowed to propagate to the next generation.

The accurate 3D phase unwrapping [39] is obtained by phase-matching algorithm proposed by ESA [55]. According to ESA [55], phase-matching algorithm is matched the phase of wrapped phase with unwrapped phase by the given equation:

$$\psi_{i,j,k} = \Delta\phi_{i,j,k} + 2\pi\rho\left[\frac{1}{2\pi}\left(\widehat{\Delta\phi}_{i,j,k} - \Delta\phi_{i,j,k}\right)\right] \tag{16}$$

where $\psi_{i,j,k}$ is the phase-matched unwrapped phase; i, j and k are the pixel positions in the quality phase map; $\Delta\varphi_{i,j,k}$ is the given wrapped phase; $\Delta\hat{\phi}_{i,j,k}$ is the approximated unwrapped phase $\rho[.]$ is a rounding function which is defined by $\rho[t] = \lfloor t + 1/2 \rfloor$ for $t \geq 0$ and $\rho[t] = \lfloor t - 1/2 \rfloor$ for $t < 0$ and i, j and k are the pixel positions in x, y and z directions, respectively [41–44].

9. Results and discussion

Figure 15 shows that the Sentinel-1A data were acquired pre-earthquake and post-earthquake on April 17 and 29, 2015, respectively. The urban zones and top of mountains are dominated with higher coherence of 0.8 and 1, respectively, as compared to vegetation and water. In contrast, the water has lower backscatter and coherence of −30 dB and 0.25, respectively (**Figure 15b**). In fact, Sentinel-1A beam mode of interferometric wide swath has spatial resolution of 5 × 20 m and swath width of 250 km with VV polarization.

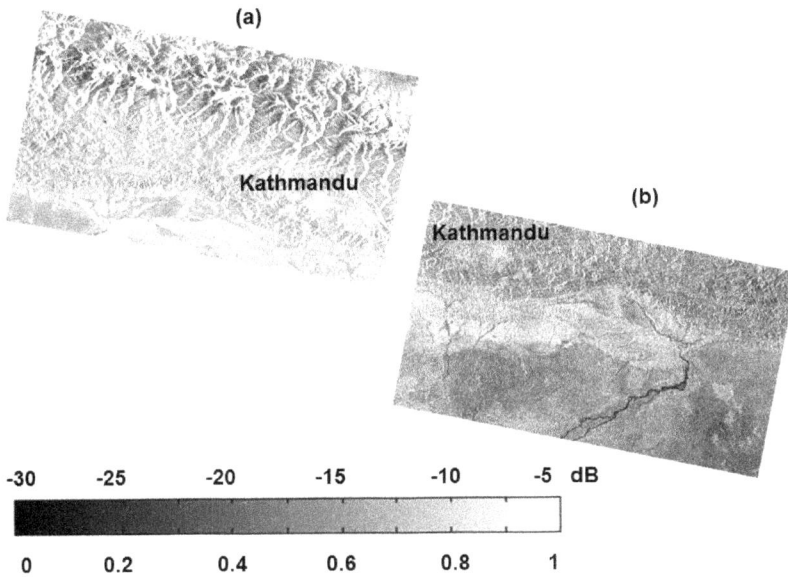

Figure 15. Sentinel-1A satellite data (a) pre- and (b) post-earthquake.

Sentinel-1 is the European Radar Observatory, representing the first new space component of the global monitoring for environment and security (GMES) satellite family, designed and developed by ESA and funded by the European Commission (EC). Sentinel-1 is composed of a constellation of two satellites, Sentinel-1A and Sentinel-1B, sharing the same orbital plane with a 180° orbital phasing difference [57]. In fact, the appropriate SAR polarization for earthquake research either VV or HH polarization modes in C-band SAR satellite data [57].

The specific needs of the four different measurement modes with respect to antenna agility require the implementation of an active phased array antenna. For each swath the antenna has to be configured to generate a beam with fixed azimuth and elevation pointing. Appropriate elevation beamforming has to be applied for range ambiguity suppression. In addition, the incident angle is ranged between 20° and 46°.

Figure 16 shows the interferogram produced by Marghany [57] by conjoining the two complex satellite Sentinel-1A SAR data which are acquired on April 17 and 29, 2015. **Figure 16** illustrates huge ground deformation by close cycle of fringe pattern which stricken due to the April 25, 2015 earthquake that smashed into Kathmandu. Furthermore, a few centimeters of ground deformation was uplifted vertically and subsided in Northern Kathmandu. Approximately, an area of 12,000 km² has been drifted also in Northern Kathmandu. In addition, 3 cm of ground deformation is represented with every fringe colour (**Figure 16**).

The interferogram fringes are produced by using the combination of three-dimensional best-path avoiding singularity loop (3DBPASL) algorithm and HGA algorithm. Clearly, the proposed algorithm for 3D phase unwrapping produced vibrant fringe cycles which indicate critical surface motion of 8.5 cm which is coincided with ground motion of 1.4 m north of Kathmandu (**Figure 17**). It is obvious that more than 1 m ground distortion is indicated by the highest concentration of fringe patterns. Moreover, the 3DBPASL shows a sharpest fringe patterns on the east-west of Kathmandu which confirms the dynamic uplift and ground subsiding across Kathmandu with an approximate value of 1.4 m.

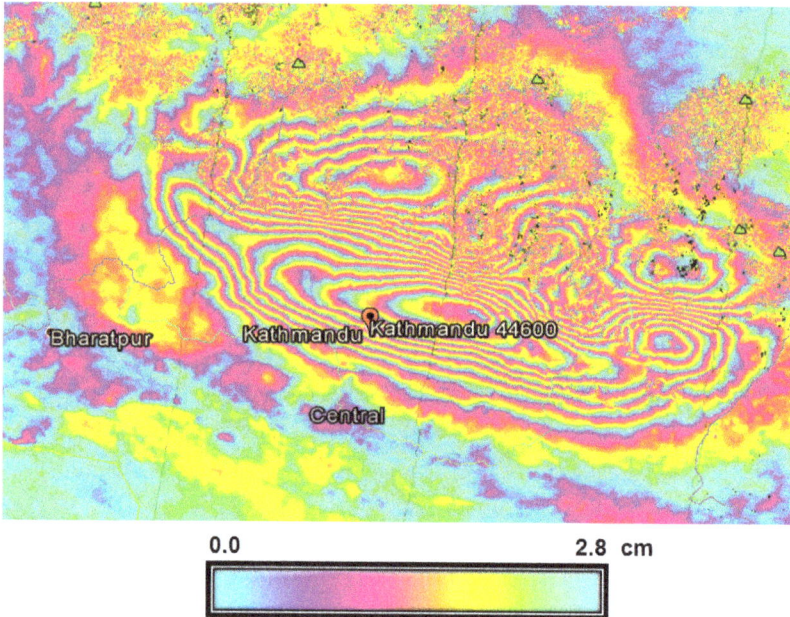

Figure 16. Sentinel-1A interferogram over Kathmandu produced by Marghany [57].

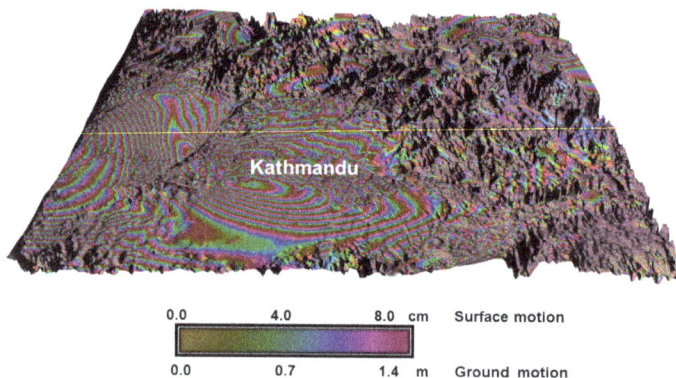

Figure 17. Interferometry produced by hybrid genetic algorithm.

This study confirms the work done by ENVISAT [19]. In fact, the 3DBPASL algorithm acquires an optimal unwrapping path, whereas the effect of singularity loops is also taken into account. In addition, zero-weighted edge is used in zero-weighted edges to adjust the optimal path and avoid these singularity loops [58].

In line with Karout [39], the quality map of phase unwrapping is precisely constructed by implementing algorithm of 3DBPASL. In fact, the proposed algorithm of 3DBPASL can determine the exact singularity loops and also remove the influences of distinctiveness loop uncertainties. Therefore, a combination of 3DBPASL for phase unwrapping with hybrid genetic algorithm produced more precisely fringe cycle. In this regard, hybrid genetic algorithm matches the phase of the wrapped phase with approximated unwrapped phase to establish the best representation of the unwrapped phase. This confirms the work done by Mughier et al. and Rajghatta [59, 60]. With this regard, a genetic algorithm is used to estimate the coefficient of an nth order of polynomial that best approximates the unwrapped phase map which minimizes the difference between the unwrapped phase gradient and the wrapped phase gradient. The genetic algorithm uses an initial solution to speed convergence. The initial solution is achieved by unwrapping using a simple unwrapping algorithm and estimating the parameters of the polynomial using weighted least squares multiple regression [44].

10. Conclusions

The deadly Nepal's earthquake was caused by an abrupt push, or due to massive stress which was built-up sideways, the foremost fault line somewhere the Indian Plate, ringing India, is gradually plunging beneath the Eurasian Plate, loud abundant of Europe and Asia. Kathmandu, consequently, located on a block of plate about 120 km wide and 60 km stretched, allegedly moved 3 m to the south in just 30 sec. The study has demonstrated new approach for monitoring the Earth deformation using advanced synthetic aperture radar sensor of Sentinel-1 A. The conventional method of InSAR based on 2D phase unwrapping has been modified to 3D phase unwrapping. This study has implemented 3D phase unwrapping based on hybrid genetic algorithm for three-dimensional Nepal's earthquake construction. In doing so, Sentinel-1A satellite data beam mode of interferometric wide swath during April 17 and 29, 2015 are used. Further, three-dimensional phase unwrapping is performed using three-dimensional best-path avoiding singularity loops (3DBPASL) algorithm. Then, phase matching is implemented with 3DBPASL using hybrid genetic algorithm (HGA). Further, polynomial surface-fitting-weighted least-square multiple regression method is implemented too as initial solution. The study shows that the top of mountain and urban areas have the highest value of backscatter of −5 dB than the surrounding environment. The study shows also that the interferogram produced by ESA has unclear pattern with 3 cm ground deformation and more than 1 m ground motion. However, 3DBPASL algorithm based on HGA provided the sharpest fringe pattern of the interferogram than ESA interferogram. In addition, the complete cycle of fringe patterns indicated 8.5 cm of the surface motion which was coincided with ground motion of 1.4 m north of Kathmandu. In conclusion, integration of the HGA with 3DBPASL phase unwrapping produces excellent 3D Nepal's earthquake surface displacement using Sentinel-1 A satellite data beam mode of interferometric. The Sentinel-1A satellite data have a great promise for disaster monitoring specially earthquake occurrence.

Author details

Maged Marghany* and Shattri Mansor

*Address all correspondence to: magedupm@hotmail.com

Geospatial Information Science Research Centre, Faculty of Engineering, Universiti Putra Malaysia, Serdang, Selangor, Malaysia

References

[1] Massonnet D and Feigl KL. Radar interferometry and its application to changes in the earth's surface. *Rev. Geophs.* (1998); 36: 441–500.

[2] Burgmann R, Rosen PA and Fielding EJ. Synthetic aperture radar interferometry to measure Earth's surface topography and its deformation. *Ann. Rev. Earth Plan. Sci.* (2000); 28: 169–209.

[3] Hanssen RF. Radar Interferometry: Data Interpretation and Error Analysis, Kluwer Academic, Dordrecht, Boston, (2001).

[4] Zebker HA, Rosen PA, and Hensley S. Atmospheric effects in interferometric synthetic aperture radar surface deformation and topographic maps. *J. Geophys. Res.* (1997); 102: 7547–7563.

[5] Hai L and Renbiao W. Robust interferometric phase estimation in InSAR via joint subspace projection. In Padron I (ed.) Recent Interferometry Applications in Topography and Astronomy, InTech—Open Access Publisher, University Campus STeP Ri, Croatia, (2012); 111–132.

[6] Askne J, Santoro M, Smith G, and Fransson JES. Multitemporal repeat-pass SAR interferometry of boreal forests. *IEEE Trans. Geosci. Remote Sens.* (2003); 41: 1540–1550.

[7] Nizalapur V, Madugundu R, and Shekhar Jha C. Coherence-based land cover classification in forested areas of Chattisgarh, central India, using environmental satellite-advanced synthetic aperture radar data. *J. Appl. Remote Sens.* (2011); 5: 059501-1–059501-6.

[8] Rao KS and Al Jassar HK. Error analysis in the digital elevation model of Kuwait desert derived from repeat pass synthetic aperture radar interferometry. *J. Appl. Remote Sens.* (2010); 4: 1–24.

[9] Rao KS, Al Jassar HK, Phalke S, Rao YS, Muller JP, and. Li Z. A study on the applicability of repeat pass SAR interferometry for generating DEMs over several Indian test sites. *Int. J. Remote Sens.* (2006); 27: 595–616.

[10] Lee H. Interferometric Synthetic Aperture Radar Coherence Imagery for Land Surface Change Detection. Ph.D theses, University of London, London, (2001).

[11] Luo X, Huang F, and Liu G. Extraction co-seismic deformation of Bam earthquake with differential SAR interferometry. *J. New Zea. Inst. Surv.* (2006); 296: 20–23.

[12] Yang J Xiong T and Peng Y. A fuzzy approach to filtering interferometric SAR data. *Int. J. Remote Sens.* (2007); 28: 1375–1382.

[13] Gens R. The influence of input parameters on SAR interferometric processing and its implication on the calibration of SAR interferometric data. *Int. J. Remote Sens.* (2000); 2: 11767–11771.

[14] Anile AM, Falcidieno B, Gallo G, Spagnuolo M, Spinello S. Modeling uncertain data with fuzzy B-splines. *Fuzzy Sets Syst.* (2000); 113: 397–410.

[15] Marghany M. Simulation of 3-D coastal spit geomorphology using differential synthetic aperture interferometry (DInSAR). In Padron I, (ed.) Recent Interferometry Applications in Topography and Astronomy, InTech—Open Access Publisher, Croatia, (2012); 83–94.

[16] Spangnolini U. 2-D phase unwrapping and instantaneous frequency estimation. *IEEE Trans. Geo. Remote Sens.* (1995); 33: 579–589.

[17] Davidson GW and Bamler R. Multiresolution phase unwrapping for SAR interferometry. *IEEE Trans. Geosci. Remote Sens. (Basel)* (1999); 37: 163–174.

[18] ENVISAT, ENVISAT Application [online]. Available from http:\www.esa.int [Accessed 2 Febraury 2016].

[19] Marghany M. DInSAR technique for three-dimensional coastal spit simulation from radarsat-1 fine mode data. *Acta Geophys.* (2013); 61: 478–493.

[20] Marghany M. Three dimensional coastal geomorphology deformation modelling using differential synthetic aperture interferometry. *Verlag Z. Naturforsch.* (2012); 67a: 419–420.

[21] Marghany M. DEM reconstruction of coastal geomorphology from DInSAR. In Murgante B. et al. (ed.) Lecture Notes in Computer Science (ICCSA 2012), Part III, LNCS 7335: 435–446. Springer Berlin Heidelberg.

[22] Marghany M. Three dimensional coastline deformation from Insar Envisat Satellite data. In Murgante B, Misra S, Carlini M, Torre CM, Quang NH, Taniar D, Apduhan BO, and Gervasi O Computational Science and Its Applications—ICCSA 2013, (2013); 7972:599–610. Springer Berlin Heidelberg.

[23] Nitecki, MH, Lemke, JL, Pullman, HW and Johnson, ME Acceptance of plate tectonic theory by geologists. *Geology* (1978); 6(11): pp.661–664.

[24] Zebker HA, Werner CL, Rosen PA and Hensley S. Accuracy of topographic maps derived from ERS-1 interferometric radar. *IEEE Geosci. Remote Sens.* (1994); 2: 823–836.

[25] Baselice F, Ferraioli G, and Pascazio V. DEM reconstruction in layover areas from SAR and auxiliary input data. *IEEE Geosci. Remote Sens. Lett.* (2009); 6: 253–257.

[26] Ferraiuolo G, Pascazio V, Schirinzi G. Maximum a posteriori estimation of height pro-files in InSAR imaging. *IEEE Geosci. Remote Sens. Lett.* (2004); 1: 66–70.

[27] Ferraiuolo G, Meglio F, Pascazio V, and Schirinzi G. DEM reconstruction accuracy in multichannel SAR interferometry. *IEEE Trans. Geosci. Remote Sens.* (2009); 47: 191–201.

[28] Ferretti A, Prati C, Rocca F. Multibaseline phase unwrapping for InSAR topography estimation. *Il Nuov Cimento* (2001); 24: 159–176

[29] Rajghatta C. Is this the 'big Himalayan Quake' we feared?. *Time India*. Retrieved 26 April 2015.

[30] Hussein A, Miguel AH, Munther G, David B, Michael L, Francis L, Christopher M, Daniel S and Mohammed Q, Robust three-dimensional best-path phase-unwrapping algorithm that avoids singularity loops. *Appl. Opt.* (2009); 23: 4582–4596.

[31] Hussein SA, Gdeisat MA, Burton DR, Lalor MJ, Lilley F, and Moore CJ. Fast and robust three-dimensional best bath phase unwrapping algorithm. *Appl. Opt.* (2007); 47: 6623–6635.

[32] Wei X and Cumming I. A region-growing algorithm for InSAR phase unwrapping. *IEEE Trans. Geosci. Remote Sens.* (1999); 37(1): 124–134.

[33] Costantini M. A novel phase unwrapping method based on network programming. *IEEE Trans. Geosci. Remote Sens.* (1998); 36(3): 813–831.

[34] Goldstein RM, Zebker HA, and Werner CL. Satellite radar interferometry: two dimen-sional phase unwrapping. *Radio Sci.* (1988); 23(4): 713–720.

[35] Ireneusz B, Stewart MP, Kampes PM, Zbigniew P and Peter L. A modification to the goldstein radar interferogram filter. IEEE Trans. Geosci. Remote Sens. (2003); 41(9): 2114–2118.

[36] Nan W, Da-Zheng F and Junxia L. A locally adaptive filter of interferometric phase images. *IEEE Geosci. Remote Sens. Lett.* (2006); 3(1): 73–77.

[37] Yu Q, Yang X, Fu S, Liu X, and Sun X. An adaptive contoured window filter for interfero-metric synthetic aperture radar. *IEEE Geosci. Remote Sens. Lett.* (2007); 4(1): 23–26.

[38] Pepe A. Advanced multitemporal phase unwrapping techniques for DInSAR analyses. In Padron I (ed.) Recent Interferometry Applications in Topography and Astronomy. InTech—Open Access Publisher, University Campus STeP Ri, Croatia, (2012); 57–82.

[39] Moreira A, Krieger G, Hajnsek I, Hounam D, Werner M, Riegger S, and Settelmeyer E, TanDEM-X: a TerraSAR-X Ad on satellite for single pass SAR interferometry, *Proc. IGARSS'05 Anchorage Alaska.* (2004); II: pp. 1000–1003.

[40] Karout S, Two-Dimensional Phase Unwrapping, Ph.D Theses, Liverpool John Moores University, England, (2007). Liverpool John Moores University, City Campus,Liverpool, United Kingdom.

[41] Haupt RL and Haupt SE, Practical Genetic Algorithms, John-Wiley & Sons, (2004).

[42] Saravana, SS, Ponnanbalam, SG, Rajendran, C. A multiobjective genetic algorithm for scheduling a flexible manufacturing system. *Int. J. Adv. Manuf. Technol.* (2003); 22:229–236. Springer Berlin Heidelberg.

[43] Schwarz O. Hybrid Phase Unwrapping in Laser Speckle Interferometry with Overlapping Windows, Shaker Verlag, (2004).

[44] Zhong H, Tang J, Zhang S, and Chen M, An improved quality-guided phase-unwrapping algorithm based on priority queue, *IEEE Geosci. Remote Sens. Lett.* (Mar. 2011); 8(2): pp. 364–368.

[45] Ghiglia DC and Pritt MD. Two-Dimensional Phase Unwrapping: Theory, Algorithm, and Software, Wiley, New York, NY, USA, (1998).

[46] Gao, D and Yin F. Mask cut optimization in two-dimensional phase unwrapping. *IEEE Geosci. Remote Sens. Lett.* (May 2012); 9(3): pp. 338–342.

[47] Falvey DA. The development of continental margins in plate tectonic theory. *APEA J* (1974); 14(1): pp.95–106.

[48] Dickinson WR Evidence for plate-tectonic regimes in the rock record. *Am. J. Sci.* (1972); 272(7): pp.551–576.

[49] Turcotte, DL, and Schubert, G Plate Tectonics Geodynamics (2 ed.), Cambridge University Press, (2002); pp. 1–21. Kuala Lumpur, Malaysia.

[50] Mitri G, Bland MT, Showman, AP, Radebaugh J, Stiles B, Lopes R, Lunine JI, Pappalardo RT Mountains on titan: modeling and observations. *J. Geophys. Res.* (2010); 115: E10.

[51] Wikipedia. List of Tectonic Plate Interactions. https://en.wikipedia.org/wiki/List_of_tectonic_plate_interactions. (2016) [Access September 25 2016].

[52] Frankel, HR. The Continental Drift Controversy: *Volume 2, Paleomagnetism and Confirmation of Drift,* Cambridge University Press, (2012). Kuala Lumpur, Malaysia.

[53] Uyeno G. and St., Writer. Buried Tectonic Plate Reveals Hidden Dinosaur-Era Sea. http://www.livescience.com/55855-newfound-tectonic-plate-east-asian-sea.html. (2016); [Access September 25 2016].

[54] Bollinger, L, Sapkota SN, Tapponnier P, Klinger Y, Rizza M, Van der Woerd J, Tiwari DR, Pandey R, Bitri A, Bes de Berc S.Return period of great Himalayan earthquakes in eastern Nepal: evidence from the Patu and Bardibas strands of the main frontal thrust. *J. Geophys. Res.* (2014);Vol (1), p. 2607.

[55] ESA. Nepal Earthquake on the Radar. https://sentinel.esa.int/web/sentinel/news/-/article/nepal-earthquake-on-the-radar.2015, [Access August 20, 2015].

[56] Marghany M. Simulation of 3-D coastal spit geomorphology using differential synthetic aperture interferometry (DInSAR). In Padron I (ed.) Recent Interferometry Applications

in Topography and Astronomy. InTech—Open Access Publisher, University Campus STeP Ri, Croatia, (2012); 83–94.

[57] Marghany M. Simulation of three-dimensional of coastal erosion using differential interferometric synthetic aperture radar. *Global NEST J.* (2014); 16 no 1: pp 80–86.

[58] Marghany, M. Hybrid genetic algorithm of interferometric synthetic aperture radar for three-dimensional coastal deformation. *Frontiers Artif. Intell. Appl. New Trends Software Method. Tools Tech.* (2014); 265: 116–131.

[59] Mughier JL, Huyghe P, Gajurel AP, Upreti BN and Jouanne F. Seismites in the Kathmandu basin and seismic hazard in central Himalaya .tif). *Tectonophysics*, (2011); 509(1–2): 33–49.

[60] Ravilious K. Nepal Quake Followed Historical Pattern http://www.bbc.com/news/science-environment-32472310. (2015); [Access on August 19 2015].

[61] USGS. Himalayan Tectonic. http://earthquake.usgs.gov/earthquakes/tectonic/images/himalaya_tsum.tif. (2015); [Access on August 19 2015].

[62] Marghany M. Three-Dimensional Hybrid Genetic Algorithm Phase Unwrapping for Nepal Earthquake Displacement Using Sentinel-1A Satellite. 36th Asian Conference on Remote Sensing: Fostering Resilient Growth in Asia, ACRS 2015; Crowne Plaza Manila Galleria, Quezon City, Metro Manila; Philippines; 24 October 2015–28 October 2015; (2015).

[63] Maître H, (ed.) Processing of Synthetic Aperture Radar Images. John Wiley & Sons, Inc, Hoboken, (2008); 384 p.

Permissions

All chapters in this book were first published in EARTHQUAKES, by InTech Open; hereby published with permission under the Creative Commons Attribution License or equivalent. Every chapter published in this book has been scrutinized by our experts. Their significance has been extensively debated. The topics covered herein carry significant findings which will fuel the growth of the discipline. They may even be implemented as practical applications or may be referred to as a beginning point for another development.

The contributors of this book come from diverse backgrounds, making this book a truly international effort. This book will bring forth new frontiers with its revolutionizing research information and detailed analysis of the nascent developments around the world.

We would like to thank all the contributing authors for lending their expertise to make the book truly unique. They have played a crucial role in the development of this book. Without their invaluable contributions this book wouldn't have been possible. They have made vital efforts to compile up to date information on the varied aspects of this subject to make this book a valuable addition to the collection of many professionals and students.

This book was conceptualized with the vision of imparting up-to-date information and advanced data in this field. To ensure the same, a matchless editorial board was set up. Every individual on the board went through rigorous rounds of assessment to prove their worth. After which they invested a large part of their time researching and compiling the most relevant data for our readers.

The editorial board has been involved in producing this book since its inception. They have spent rigorous hours researching and exploring the diverse topics which have resulted in the successful publishing of this book. They have passed on their knowledge of decades through this book. To expedite this challenging task, the publisher supported the team at every step. A small team of assistant editors was also appointed to further simplify the editing procedure and attain best results for the readers.

Apart from the editorial board, the designing team has also invested a significant amount of their time in understanding the subject and creating the most relevant covers. They scrutinized every image to scout for the most suitable representation of the subject and create an appropriate cover for the book.

The publishing team has been an ardent support to the editorial, designing and production team. Their endless efforts to recruit the best for this project, has resulted in the accomplishment of this book. They are a veteran in the field of academics and their pool of knowledge is as vast as their experience in printing. Their expertise and guidance has proved useful at every step. Their uncompromising quality standards have made this book an exceptional effort. Their encouragement from time to time has been an inspiration for everyone.

The publisher and the editorial board hope that this book will prove to be a valuable piece of knowledge for researchers, students, practitioners and scholars across the globe.

List of Contributors

Telman Aliev
Institute of Control Systems of the Azerbaijan National Academy of Sciences, Baku, Azerbaijan

Tamás János Katona
University of Pécs, Pécs, Hungary
MVM Nuclear Power Plant Paks Ltd., Paks, Hungary

Yolanda Alberto-Hernandez and Ikuo Towhata
University of Tokyo, Japan

Mustafa Toker
Department of Geophysical Engineering, Yuzuncu Yıl University, Van, Turkey, Sodankylä Geophysical Observatory (SGO), University of Oulu, Oulu, Finland

Zhan Wen, Xiao-hui Zeng and Feng Zou
School of Communication Engineering, Chengdu University of Information Technology, Chengdu, China

Ya-juan Xue
School of Communication Engineering, Chengdu University of Information Technology, Chengdu, China
School of Geophysics, Chengdu University of Technology, Chengdu, China

Jun-xing Cao
School of Geophysics, Chengdu University of Technology, Chengdu, China

Gu-lan Zhang
Bgp, Cnpc, Zhuozhou, Hebei, China

Hao-kun Du
Geophysical Institute of Zhongyuan Oil Field, Sinopec, Henan, China

Jae Cheon Jung
Kepco International Nuclear Graduate School, Seosang-myeon, Ulju-gun, Ulsan, Republic of Korea

Index

www.ingramcontent.com/pod-product-compliance
Lightning Source LLC
Chambersburg PA
CBHW062001190326
41458CB00009B/2930